Green Manufacturing

T0140320

David A. Dornfeld

Editor

Green Manufacturing

Fundamentals and Applications

 Springer

Editor
David A. Dornfeld
Laboratory for Manufacturing and Sustainability (LMAS)
University of California, Berkeley
Berkeley, California, USA

ISBN 978-1-4899-9015-0   ISBN 978-1-4419-6016-0 (eBook)
DOI 10.1007/978-1-4419-6016-0
Springer New York Heidelberg Dordrecht London

Printed on acid-free paper

Springer is part of Springer Science+Business Media (www.springer.com)

Preface

The interest and enthusiasm around the topic of green manufacturing, as a first step towards sustainable manufacturing, has encouraged a lot of research and investigation into this complex subject This book tries to give perspective to the term green manufacturing—how it is defined, where is "fits in" relative to sustainable production, what are the basic guidelines and tools of the trade for analyzing and practicing green manufacturing, and some examples of applications. The content reflects the research activities of the Laboratory for Manufacturing and Sustainability at the University of California, Berkeley, and is presented in the voice of the number of student researchers in the lab over the last few years. This work was motivated by the relatively little material on this topic in publication but the increasing interest from the public and our industrial and institutional partners. We are also motivated by the importance of this aspect of manufacturing, our curiosity, and the need for educational material in graduate course offerings in Mechanical Engineering at Berkeley and elsewhere.

As this is being written, research in our lab and many other labs around the world continues on subjects related to green manufacturing. There is undoubtedly much more that could be included here but that will not be due to the continuously evolving nature of the subject and the constant advance of research and understanding. Like painting the Golden Gate Bridge across San Francisco Bay, when one finishes at one end it is time to start again from the beginning!

The perspective presented in this book is from the level of the manufacturing process, machine, and system, as well as the supply chain and packaging, since for many companies the bulk of the impact comes from items or processes outside of their immediate control. There is a lot of excellent strategic discussion about the importance of green and sustainable manufacturing. But, at the execution level, there is little engineering information—the material practicing engineers need to fulfill the higher aspirations of their organizations as they move towards sustainability. The authors hope that this view into green manufacturing from our perspective will help the reader to understand a bit more some of the practical aspects of the topic and encourage them to look more closely at the processes and systems around them in their work or research to observe the opportunities for improvement, replacement, reuse, and, overall, reduction of impact. It should also allow them to quantify the effects of these improvements and impact reductions and, hence, support the overall goal of greening manufacturing as part of the move towards sustainable manufacturing.

This book is intended to be used as a basic reference text for industrial practitioners to get up to speed on green manufacturing as well as a basic textbook for an advanced course complimented by readers from some of the books we refer to in this compilation. We will be using it in the graduate course in Mechanical Engineering on Sustainable Manufacturing at UC Berkeley. Your feedback and comments on the book will be appreciated.

A lot of the interest in the topic of green manufacturing at Berkeley has been generated by readers of the green manufacturing blog—green-manufacturing. blogspot.com. If you have seen this blog you will see a similar tone of discussion. We intend to keep the discussion going on the blog to extend the impact of this book and allow the incorporation of new material as it appears.

Finally, most of the credit for the content of this book goes to the graduate student researchers and visiting researchers to the Laboratory for Manufacturing and Sustainability. Those who have specifically contributed to this work are listed at the beginning of each chapter. Others who contributed based on their past work are cited. Current research information and student and visiting researcher projects can be found on our lab Web site lmas.berkeley.edu. We also appreciate the very capable assistance of Ms. Lexie Cousens who provided editorial assistance in the production of this book.

Berkeley, CA, USA   David Dornfeld

Contents

Introduction to Green Manufacturing

1

David Dornfeld, Chris Yuan, Nancy Diaz, Teresa Zhang,
and Athulan Vijayaraghavan

> *If you are not part of the solution, you are part of the problem.*
>
> Eldridge Cleaver

Abstract

This chapter has as its objective a basic introduction to the topic to set the stage
for the rest of the book. It introduces first the importance of this topic now and
then the motivation, basics, and definitions associated with green manufacturing
and sustainability. It describes some of the drivers that are causing governments
and industry to take steps to green their processes, machines, systems, and

D. Dornfeld (✉)
Laboratory for Manufacturing and Sustainability (LMAS), Department of Mechanical
Engineering, University of California at Berkeley, 5100A Etcheverry Hall, Mailstop 1740,
Berkeley, CA 94720-1740, USA
e-mail: Dornfeld@berkeley.edu

C. Yuan
University of Wisconsin, Milwaukee, WI, USA
e-mail: cyuan@uwm.edu

N. Diaz
LMAS, Mechanical Engineering Department, University of California, Berkeley, CA 94720-1740,
USA
e-mail: ndiaz08@gmail.com

T. Zhang
R.W. Beck/SAIC Energy, Environment & Infrastructure, 1671 Dell Avenue, Suite 100, Campbell,
CA 95008, USA
e-mail: TERESA.W.ZHANG@saic.com

A. Vijayaraghavan
System Insights, 2560 Ninth Street, Suite 123A, Berkeley, CA 94710, USA
e-mail: athulan@systeminsights.com

D.A. Dornfeld (ed.), *Green Manufacturing: Fundamentals and Applications*,
DOI 10.1007/978-1-4419-6016-0_1, © Springer Science+Business Media New York 2013

1

enterprises. A discussion about the distinction between green and sustainable is introduced with respect to incremental improvements, greening and achieving overall sustainability. Strategies for achieving green manufacturing are presented. Barriers and obstacles to greening manufacturing are presented along with examples from industrial practice.

1.1   Why Green Manufacturing?

1.1.1   Background

It is important to define, as best as one is able to, why it is important to focus on green manufacturing. Green manufacturing is an important part of business. And business must be analyzed "holistically"—that is, do not fiddle with just little parts. Do it right for the whole system. Here's what Paul Hawken and the Lovins said in Natural Capital [1]:

> Without a fundamental rethinking of the structure and the reward system of commerce, narrowly focused eco-efficiency could be a disaster for the environment by overwhelming resource savings with even larger growth in production of the wrong materials, in the wrong place, at the wrong scale, and delivered using the wrong business models.

They go on to say that the best solutions are not simply some kind of tradeoff between the objectives of economic, environmental, and social capital (the three pillars of sustainability) but an integrated approach at all levels from "technical devices to production systems to companies to economic sectors to entire cities and societies."

Sounds like manufacturing is a big part of that. There are strong business pressures to embrace green manufacturing as will be discussed below. But, there is more.

All the data on energy consumption, global temperatures, CO_2 levels in the atmosphere, other impacts of industrialization, and population growth head up and to the right in the graphs—meaning things are moving toward more challenging, and less sustainable, conditions. One may or may not fully agree with the predictions, but from the perspective of cost of energy, availability of energy, cost of treatment/disposal of waste products, materials, etc., things will get more expensive. And, legislation marches on. The recent deliberations in the United States Congress have a goal to reduce CO_2 by 83% by 2050 and envision some form of cap and trade program [2]. Governments in other parts of the world have even more aggressive plans for reduction. If one relies on a global market for one's products, as increasing numbers of organizations do, one will be faced with the demands of green manufacturing sooner or later.

Regardless of an individual's feelings towards the severity of the situation, there are forces moving to make it more and more expensive and difficult to continue "business as usual" with respect to environmental impacts of manufacturing around the world. Individuals and organizations should be prepared and use this to their advantage.

Consider where something, say an auto, is manufactured. One can make a simple analysis of energy needed to make an automobile. This is called "embodied energy" and expressed in units of kWh and is an estimation of the total energy needed to make the car, not to operate it—the so-called manufacturing phase not use phase. Then, through the magic of conversions, an estimate the greenhouse gases (GHG) attributable to that embodied energy can be determined by converting from kWh to GHG using factors that are based on the source of the electricity; that is, from coal or other carbon-based energy sources, or hydro, solar or wind and other renewable sources, or nuclear. Carbon-based energy has a higher GHG impact than renewable. Although there are other measures of impact, green house gas offers a convenient and widely used proxy. This allows a clear view of the impact of where we manufacture something. Interestingly, making the same vehicle in different places (depending on the energy mix) will result in dramatically different GHG output. The data on which this is based is found on the US DOE Web site [3] or on the Get Energy Active Web site [4].

Let's look at some examples. If one builds a "typical auto" in France, which has most of its electrical energy generated in nuclear plants, the GHG impact of manufacturing that auto will be about one-seventh that of building it in the USA, or less than one-tenth that of building it in China. Zooming in on the USA, there are great differences between states also. Building a car in California with its mix of renewable energy vs. Kentucky with its dependence on coal fired plants means a factor of 4 difference in GHG impact between the same manufacturing processes—only based on location. This is for manufacturing the same car, same features, same performance—only accounting for the location at which the energy is consumed in manufacturing.

Granted, there are some simplifications in the above example. Few, if any, automobiles are produced entirely at one location since they rely on a complex supply chain. But, the effect of location is impressive.

So why does this matter? Regulations already exist that apply penalties for excessive GHG emission in products during the use phase. For example, if a consumer wants to buy an auto in France they will see listed, along with fuel consumption in liters/km, the GHG generated in units of grams of CO_2/km traveled. Similar information is available for Ford in the UK. The CO_2 emission is listed in g/km and the performance can be compared over a range of engines [5]. And, if a vehicle is purchased with a large engine that emits GHG above a certain level—the consumer pays more. If the customer purchases one with emission below a certain level, the cost is less. Same manufacturer, same quality vehicle, same operation, same manufacturing process—but costs more if it emits more GHG in operation. This will happen in due time for manufacturing based on embedded energy in the product due to manufacture (that is, embedded energy in the materials, water, energy, consumables, etc.).

If a company's product is a machine tool their customer will worry about the energy and resources used to build the machine tool (because it may affect the cost). And then, when it is installed in the factory, the customer will be worried about how much energy and other resources the machine consumes in its "use phase." And then there is the transportation cost from the manufacturing site to the distribution site and customer. All those impacts should be included as well.

And so it goes! The smart manufacturer will optimize where to build the product in terms of energy mix and transportation costs to the consumer. One should probably add availability of water and the energy cost of providing that as well. These kinds of considerations will, in our opinion, greatly influence the location of manufacturing (and, hopefully, offset some of the fascination with low labor costs as the sole determiner of location).

Paraphrasing Lord Kelvin, "If you can't measure what you make, you don't know if you've made it or not." We need to be able to understand and measure the resources used on our products and their use. Then we can make informed decisions about their design, distribution, and utilization. This really encourages one to think about the life cycle costs of energy and consumables in the manufacture of a product—an important driver for green manufacturing.

1.1.2 How Do We Define Green?

"Green" is an adjective that, in the context of this book, is defined as "concerned with or supporting environmentalism and tending to preserve environmental quality (as by being recyclable, biodegradable, or nonpolluting)" [6]. This definition alone is broad, but when applied to manufacturing the general idea of green manufacturing is a process or system which has a minimal, nonexistent, or negative impact on the environment. A definition adapted from one proposed by the U.S. Department of Commerce has sustainable manufacturing as "the creation of manufacturing products that use materials and processes that minimize negative environmental impacts, conserve energy and natural resources, are safe for employees, communities, and consumers and are economically sound" [7].

The term "green" can also be used as a verb which would then refer to the process of reducing the environmental impact of a manufacturing process or system when compared to a previous state. Examples of greening a manufacturing system therefore include reducing the volume of hazardous waste produced, cutting down on the coolant consumed while machining, even changing the energy mix to include more renewable energy sources.

An important distinction to be made is that between green manufacturing and lean manufacturing which some use interchangeably, even though the two systems have different end goals. Lean is centered around creating more value *for the customer* with less work. Toyota was a key player in the development of lean manufacturing and in the process of doing so Toyota targeted the reduction of seven wastes as part of the Toyota Production System, or TPS, as [8]:

- Overproduction (including early production)
- Transportation (and handling including delays, handling several times, unnecessary moving or handling)
- Inventory (including work in process, buffers, and finished stock)
- Motion (unnecessary non value added actions of people or machines)

- Defects (including making excess scrap and need for rework/repair)
- Over-processing (unnecessary production steps of procedures that add no value to the product)
- Waiting (any idle time or unnecessary non-value added time)

Now, several of these wastes can be related to the desire to minimize the environmental impact of a process. For example, if waiting time were decreased, one may argue that resources such as factory lighting and air conditioning would be used more efficiently. Many production machines use a lot of energy even when they are not producing any product or processing. So, idle time designed to allow smooth flow of product at other machines wastes energy. Another example relating a lean system to a green system would be in the reduction of transportation distances of a product within the factory walls which potentially decreases the time required to transport the product, while reducing the energy consumed in the transportation process. Related to making transportation more efficient though is the introduction of alternatives like a conveyor belt to replace the transportation of a product that would have otherwise been done by physical labor. Although transporting a product by means of a conveyor belt would mean that the product may arrive more rapidly at the next station and would therefore decrease wait time, the carbon emissions per meter traveled associated with transporting by conveyor belt are much greater than that compared to physical labor.

Lean manufacturing usually refers to practice that attempts to identify and evaluate the use of resources for anything other than adding value for the end customer. It then defines these uses as wasteful and tries to eliminate them. Further to this, it is understood that there are a number of approaches to "lean." The first approach is nominally the elimination of waste and the tools that assist in uncovering waste in the process and system and getting rid of it. A second approach, more aligned with the Toyota Production System, focuses on the "smoothness" of production and constructing a process with the capability to produce the required results by designing out process inconsistency (or "muri"). This is to be done while trying to maintain as much flexibility as possible since excessive constraints or rigidity often induce waste (as in excessive set up/change over time, high minimum run lot sizes requiring extra inventory or inducing poor response to customer needs, i.e., poor response to "pull"). So, lean manufacturing concepts include many useful aspects for making a manufacturing process green but may not consider all elements. Certainly, it is a helpful start.

In contrast to lean manufacturing, environmentally benign (or conscious) manufacturing is much closer to green manufacturing in that it "addresses the dilemma of maintaining a progressive worldwide economy without continuing to damage our environment." The focus is therefore shifted from concentrating on the customer values to the effect on the environment [9].

1.1.3   Is Green Sustainable?

How does one distinguish between green manufacturing and sustainable manufacturing? There are a number of definitions of sustainable and sustainability.

Fig. 1.1 Manufacturing in
relation to the three pillars
of sustainability

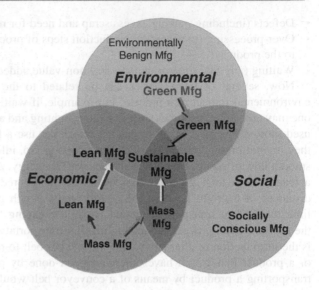

Sustainable development has been defined by the World Commission on Environment and Development as "development that meets the needs of the present without compromising the ability of future generations to meet their own needs." [10]. This definition includes within it two key concepts:

1. "The concept of 'needs,' in particular the essential needs of the world's poor, to which overriding priority should be given.
2. The idea of limitations imposed by the state of technology and social organization on the environment's ability to meet present and future needs."

Reflecting this definition to include social assessments along with economic and environmental assessment, sustainability has also been associated with the triple-bottom line approach to thinking where environmental, economic, and social factors have to be met. This can be related to manufacturing as in Fig. 1.1. Sustainable manufacturing refers to a manufacturing system or process that meets these three critical factors—that is addresses impacts on the environment, on the economy, and on society. Recall the U.S. Department of Commerce's definition above of sustainable manufacturing including "minimize negative environmental impacts, conserve energy and natural resources, are safe for employees, communities, and consumers and are economically sound"—the triple bottom line! A company could not survive if it was, at the minimum, not breaking even with respect to its finances (that is, not financially sustainable).

The social factor also plays an important role in manufacturing as is evident with the introduction of fair trade regulations and the move away from child labor practices, to name a few examples. Thus this pillar is also critical to the success of manufacturing firms but us often the most difficult to define in tangible terms that relate to business practice or manufacturing. Other measures of social impact can include worker training levels, pay levels, employee retention, job-related accidents or injuries, and so on. Lastly, with growing economies and therefore growing

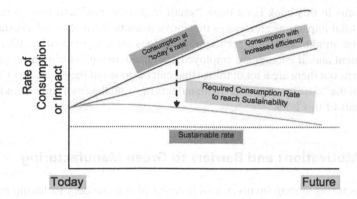

Fig. 1.2 Illustration of consumption and impact over time

demand for production, greenhouse gas emissions and resource use has become an inherent problem that leads to the last factor, the environment. It is typical for companies to achieve two out of the three factors. Achieving success in all three factors, though, is extremely difficult, but ideal. The extent to which companies are achieving the "triple bottom line" is often reflected in their corporate sustainability reports (CSR) published on their Web sites.

The premise in this book is that green (and greening manufacturing) is a step towards sustainable manufacturing. Sustainability is a specific term with a specific meaning. According to Graedel and Howard-Grenville [11] "a crucial important property of sustainability is that the concept is an absolute, as are pregnant and unique, to use two common examples. A sustainable world is not one that is slightly more environmentally responsible than it was yesterday." One way to visualize the situation is too look at the history of consumption or impact and the direction it is taking in the future, Fig. 1.2.

A sustainable rate of consumption of a resource, or impact due to some process or operation can be defined following the definitions above. For example, in California the sustainable rate of consuming water is determined by rainfall received during the winter months which accumulate in ground water and Sierra snowpack which melts during the warmer months. It is clear when more than a replenishable amount of water is being consumed. Similarly, the environment can accommodate certain levels of impact, greenhouse gases, for example, due to the natural processes of the earth's ecosystem. Scientists, politicians, and others are now discussing how much greenhouse gas can be exhausted to the atmosphere to avoid long term effects. Most agree that there is already more carbon dioxide being input into the atmosphere than can be naturally accommodated—so the rate of consumption (and the impact to the atmosphere) is not sustainable. Small improvements that help to reduce consumption or impact will have some effect but, unless bundled with a lot of others, will never achieve sustainability.

The focus in this book is on these "small improvements" and how to measure their potential impacts as green steps that move manufacturing toward a sustainable world. One might call them technology wedges that, individually, offer some improvement and, if enough are employed, can make manufacturing more sustainable. It turns out there are a lot of things that can be done but they must be evaluated in terms of the "cost" vs. "benefit" to manufacturing and the environment and that is a major part of this book.

1.2 Motivations and Barriers to Green Manufacturing

A manufacturing system involves a wide range of stakeholders including material and other suppliers, manufacturers, retailers, consumers, and policy-makers. As the stakeholders are becoming more aware of the values of green manufacturing in practice, the manufacturing industry is motivated to implement green manufacturing strategies to reduce the environmental impact and moreover, to improve the economic performance of its manufacturing operations, as indicated by a number of research results [12–15].

1.2.1 Motivations for Green Manufacturing

There are a number of drivers in effect motivating the manufacturing industry towards green manufacturing practices. The motivation factors are summarized as mainly in the three categories of regulatory pressure, economic incentives, and competitive advantages. Specific examples of these motivations include the following:

- Pressure from Government—Regulations, Penalties, and Tax benefits
- Interest in Efficiency/Reduced Cost of Ownership (CoO)
- Scarcity of Resources/Risk
- Continuous Improvement
- Pressure from Society/Consumers/Customers and other Competitors
- Desire to Maintain Market Leadership
- Insure Control of Supply Chain Effects (what's happening outside of your facility)

Several of these categories will be looked at in more detail.

1.2.1.1 Regulatory Pressure

Left alone, populations, groups, organizations, corporations often fall into behavior patterns that threaten some aspect of the health or well-being of the group over time. This tendency has been aptly described by a scenario referred to as "tragedy of the commons" first sketched out by [16]. This scenario describes a community sharing a common green area in the center of the village (the commons.) It is used by all to pasture their animals and, if everyone respects the right of the others to equal access of grass for their animals everyone's animal gets fed and all benefit. Over time, some add additional animals of their own since they are aware that if they feed more animals they can personally benefit by selling the extra product. The upper limit on this is, of course, the ability of the commons to feed a fixed number of animals. More than the maximum and the commons deteriorates due to over feeding and every one suffers as the grass dies, or cannot be replenished, due to over feeding. A similar phenomenon occurs with over fishing certain bodies of water. In such situations, the towns folks can either "self-regulate" (which requires cooperation among all) or make rules governing all event though only one or two abused the system, hence, the motivation for regulations.

Aware of the significant environmental problems of industrial wastes and emissions, governmental agencies initiated the efforts for environmental impact control and restoration by making a series of policies, regulations, and laws, which has achieved significant progress in advancing the environmental performance of industrial production activities. U.S. environmental regulations have undergone three stages since 1970 [17, 18]. The first stage focused on the end-of-the-pipe control of environmental wastes. Representative regulations include the Clean Air Act, the Clean Water Act, and the Resource Conservation and Recovery Act (RCRA). The second stage focused on reducing the environmental pollution of industrial activities, with the Pollution Prevention Act enacted in 1990. The third stage focuses on clean production by encouraging implementation of comprehensive environmental programs to reduce the overall impact of industrial production.

Besides the above comprehensive regulations and laws, national programs on management of specific types of emissions are also established to facilitate the effective control of environmental wastes and emissions from industrial production activities. For example, the Toxic Release Inventory (TRI) program established by the US EPA requires manufacturing and seven related industries to report their toxic chemical releases annually so as to encourage and monitor the industrial efforts in reducing the amount of toxic materials being released into the environment. Regarding the intensive carbon emissions of industrial activities, the U.S. Department of Energy has established a voluntary greenhouse gas registry program for industrial manufacturers [19]. In response, large manufacturing companies such as Ford, GM, HP, and Honda. have put dedicated efforts into greenhouse gas reporting and reduction through this program.

Besides these national regulations and laws, the federal government and international organizations also put effort in organizing green manufacturing themed conferences and workshops to motivate the industry in implementing sustainability strategies in industrial production. For example, the U.S. Department of Commerce has organized a "green manufacturing day" annually since 2007 to "enhance public and private interactions in the field of green manufacturing" [7]. The Organization for Economic Co-operation and Development (OECD) organizes international conferences and workshops on green manufacturing every year to help its member countries advance their green manufacturing initiatives [20, 21].

1.2.1.2 Economic Incentives

Under the regulatory pressure and governmental efforts, the manufacturing industry is driven towards green manufacturing by the economic benefits which could result from the implementation of sustainability programs. Green manufacturing, in general, includes such practices as pollution prevention, product stewardship, and emission control [22, 23]. The economic cost involved in emission control is tremendous for the manufacturing industry. It has been reported that the U.S. manufacturers spend approximately $170 billion per year in waste treatment and disposal costs [24]. Appropriate green manufacturing programs such as pollution prevention for minimizing waste generation in manufacturing could effectively cut the costs on both waste management and material consumption, and accordingly can improve the profit margin of the manufacturing industry. A recent survey on the U.S. commercial carpet manufacturers indicates that 84.6% of the manufacturers that adopt emission control strategies such as recycling water and diverting solid waste from landfills, and 100% of the manufacturers that adopt pollution prevention strategies like reducing raw materials usage and energy consumptions have success-fully decreased their manufacturing cost [23]. More recent examples of savings due to decreased material, energy or other resource consumption are reported in the press daily. These are often documented in corporate sustainability reports or other industry documents.

1.2.1.3 Competitive Advantages

The third driver for green manufacturing is the competitive advantages a company could obtain in the market through the implementation of green manufacturing strategies. This is driven by the consumer him or herself, other commercial consumers (other industries) trying to green their processes and a desire to maintain market leadership. Society is becoming more aware of the environmental issues of manufacturing, and the customers as well as many other stakeholders typically give their preferences to those manufacturers with a better environmental image in the market. Due to the public's awareness and concerns, implementing green manufacturing can enhance a company's and an industry's image, and also can generate more revenues and increase market share [25]. A survey of 1,000 U.S. manufacturers found that 90% have environmental strategies and 80% have an environmental-friendly operations mechanism [26]. A recent analysis of over 4,000 manufacturing facilities in seven OECD countries found that there is a positive correlation between environmental performance and commercial success [27]. With such economic incentives, green manufacturing has been emerging as a new competitive requirement and a means to achieve differentiation in the market [28, 29].

In addition to the drivers detailed above the concern about reducing risk in the supply chain of any of the materials and consumables needed in manufacturing is a constant issue. Water, which in many parts of the world can have more variability in supply than energy, is now an important consideration in locating companies and production control. As this book is being written there are concerns about the availability of some rare metals due to political disputes totally unrelated to economic cycles or market forces. Green manufacturing analyses and solutions must increasingly address these challenges.

1.2.2 Barriers to Green Manufacturing

Although green manufacturing is driven by a number of positive factors, the manufacturing industry still faces some barriers and challenges that hinder the application of green manufacturing strategies in practice. Some early investigations have stated that environmental initiatives may induce a negative impact on company performance [30, 31]. In general, the barriers for green manufacturing could be summarized into the following three categories: economic barrier, technological barrier, and managerial barrier.

Early green manufacturing practices focused a lot on the emission control and waste management [24]. In the emission control and waste management process, the capital cost requirements were high and may take a long time to be paid back. In some circumstances, the capital input of emission control may exceed the total amount of direct economic gains. This has greatly hindered the practical applications of green manufacturing strategies in industry. But as green manufacturing practices are switching from end-of-the-pipe emission control to pollution prevention and the costs for waste disposal and environmental emissions are increasing, the economic barrier of green manufacturing is gradually diminishing for the manufacturing industry.

Another major barrier for green manufacturing is that the manufacturing industry has to rely on certain processes, technologies, or materials to make its products, which may cause undesirable effects but which cannot be avoided in the current stage due to the lack of appropriate technologies or processes. Take automotive manufacturing as an example, the painting operations generate a significant amount of volatile organic compound (VOC) emissions which cause air pollution by creating ozone and carcinogens. It has been reported that approximately 80% of Ford's toxic pollutants that were released into the environment were from the painting operations [32]. Even with such an enormous impact identified, complete elimination of the painting emissions is not practical in this stage as the industry lacks appropriate technologies to replace the process. Efforts to minimize the impacts are underway in most major industries. For example, in automotive painting attempts are being made to reduce the need for undercoat or primer coats and to use other paint formulations and vapor recovery.

The other major barrier for green manufacturing is that the manufacturing industry lacks capable scientifically-based decision support tools for effective implementation of green manufacturing strategies. To achieve green manufacturing, the industry needs appropriate analytical tools to characterize and benchmark the environmental impact of emissions and wastes from a specific manufacturing process/system to support decision-making. While manufacturing is such a complicated system that numerous types of processes, materials, and system patterns are employed, generic decision tools are difficult to use for the whole manufacturing industry as each manufacturing process/system has its own specificities. Besides, the manufacturing system is closely linked to many other industrial activities and the products from manufacturing impact almost everyone in society. As a result,the environmental impact of manufacturing must be assessed both comprehensively and specifically for robust decision support in the industrial applications. This needs further research efforts on environmental impact assessment methods and manufacturing process modeling and characterizations. For example, life cycle impact assessment, as a comprehensive system tool for environmental impact analysis, needs to be standardized, streamlined, and further improved before wider application in industrial practice is seen [33]. And, design and process selection tools are needed that include environmental and other sustainability impact analyses so the designer and process engineer can address issues arising from design and processing at an early stage.

1.3   Environmental Impact of Manufacturing

Manufacturing is both material- and energy-intensive. It usually consumers water also. Environmental impacts of manufacturing result mainly from the materials, water, and energy consumed in the manufacturing systems. Manufacturing is dominant in its environmental impacts in such categories as toxic chemicals, waste generation, energy consumption, and carbon emissions [34]. In the following sections, the significance of the environmental impact of manufacturing is elaborated in more detail from the perspective of toxic chemical release, waste generation, energy consumption, and carbon emissions.

1.3.1 Toxic Chemical Releases

Toxic materials are widely and heavily used in many manufacturing industries for both product development and process operations. There are concerns due to their toxic effects and significant impact on the environment and human health. There is a wide variety of toxic chemicals involved in various manufacturing operations for etching, forming, catalyzing, cleaning, etc., and such chemicals inevitably lead to waste/emissions generation from the process operations. Aware of the significant impact of toxic chemicals released from the manufacturing industry, the US EPA established the Toxic Release Inventory (TRI) program in 1987 to collect the facility level of toxic chemical release information, initially from manufacturing. In 1998, this was expanded to seven other related industrial sectors including metal mining, coal mining, electric utilities, petroleum bulk terminals, chemicals wholesalers, RCRA commercial hazardous waste treatment, and solvent recovery. According to the TRI statistics, a huge amount of toxic chemicals is released into the environment annually from both manufacturing and the seven other industrial sectors, which causes a significant impact on the public health and the eco-systems. Figure 1.3 shows the amount of toxic chemical releases from manufacturing and seven industrial sectors in the United States in the year 2001 alone (here the releases

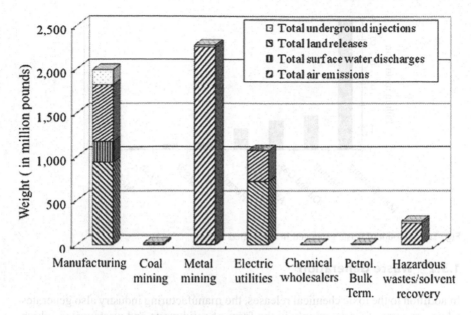

Fig. 1.3 The U.S. toxic release inventory in 2001

from RCRA commercial hazardous waste treatment and solvent recovery are combined together) [35].

These toxic chemicals, based on their release patterns, are generally categorized into four groups: air emissions, surface water discharges, land releases, and underground injections. From the data reported in the TRI, in 2001 the U.S. manufacturing industry released a total of 1.99 billion pounds of toxic chemicals into the environment, and the 2.3 billion pounds of toxic releases from metal mining can also be considered as an upstream release of the manufacturing industry, which generates significant impact on the environment and human health after being released into nature. Based on weight, the air emissions roughly account for 30% of the total toxic release; the total land releases take roughly 62% share; surface water discharges and underground injections are both around 4%. It should be noted here that the total TRI amount is an underestimate of toxic chemical releases in the United States since many small scale manufacturing entities are not required to report their toxic chemical emissions. As a result, the actual environmental and human health impacts resulting from toxic chemical release of the manufacturing industry are much more significant than what has been reflected in the TRI database.

Besides the United States, such inventory programs for collecting toxic chemical information from industrial emissions have also been established in many other countries including European Union nations, Australia, Canada, Japan, and Korea [36, 37], for monitoring and controlling the toxic chemicals released into the environment.

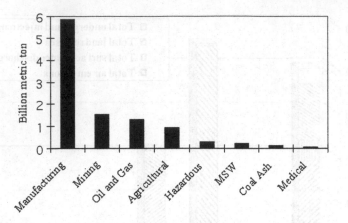

Fig. 1.4 Amount of waste generated in the United States

1.3.2 Waste Generation

In addition to the toxic chemical releases, the manufacturing industry also generates a huge amount of waste, mainly in the form of solid waste and waste water, which also causes significant environmental concerns and impacts. As reported, waste generated in the United States is more than that generated by any other single country in the world, in both absolute scale and per capita [34, 38].

Figure 1.4 shows the amount of waste generated by manufacturing and seven common industries [39]. As demonstrated in Fig. 1.4, the manufacturing industry generates much more waste than any other of these seven industrial sectors. The amount of waste generated from manufacturing is even larger than the sum of all other seven industries combined. Wastes generated from manufacturing cause significant environmental problems. Reducing the waste generation from the manufacturing industry to reduce its environmental impact is one of the major objectives of green manufacturing research and practice.

1.3.3 Energy Consumption

The environmental impact of manufacturing also comes from the enormous amount of energy consumed in the manufacturing processes and systems. The manufacturing industry is very energy-intensive. As an indication of the scale, in 2003, the manufacturing industry as a total consumed approximately 23% of total energy in the United States [40]. Figure 1.5 shows the percentage of energy consumption of U.S. economic sectors, in which the industrial sectors are broken down to manufacturing and nonmanufacturing sectors [40]. The energy consumption of manufacturing, as shown by the statistical data, is only second to that of transportation but larger than that of both residential and commercial sectors. And, in many cases transportation is a major element in manufacturing due to the complex supply chains employed.

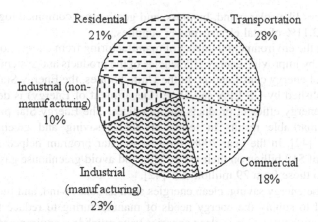

Fig. 1.5 U.S. energy consumption by economic sectors

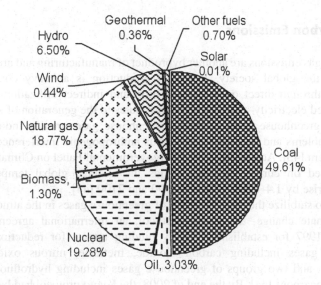

Fig. 1.6 U.S. electricity mix

Energy consumption causes environmental impact due to the fact that current energy is mainly supplied from fossil fuels. Since fossil fuels contain various polluting elements such as carbon, sulfur, and nitrogen, the electricity generated from fossil fuels produces significant amount of pollutants including carbon dioxide, sulfur dioxide, and nitrogen oxides, which cause significant environmental problems such as global warming, acidification, and smog.

In Fig. 1.6, U.S. electricity generation is shown by energy source by type of producer in the year 2006 [41]. For the total amount of 4,065 TWH electricity generated in 2006 in the United States, 71.4% is generated from fossil fuels. Clean energy sources like solar, wind, biomass, and geothermal combined together only account for 2.11% of total electric power supplies.

Reducing the environmental impact of manufacturing from energy consumption can be made by improving energy efficiencies of the products and systems to reduce the amount of energy consumption. In the United States, the Energy Star program, jointly established by US EPA and DOE (Department of Energy) is dedicated to developing energy efficient products and practices. The Energy Star program has achieved remarkable results in terms of energy-saving and greenhouse gas eliminations [42]. In the year 2008, the Energy Star program helped Americans save a total of $19 billion on their utility bills and avoid greenhouse gas emissions equivalent to those from 29 million cars [42].

Besides the energy saving, clean energies such as solar, wind, and fuel cells can also be used to supply the energy needs of manufacturing to reduce the energy impact of manufacturing since clean energies have much less environmental impact than the electricity generated from fossil fuels.

1.3.4 Carbon Emissions

Greenhouse gas emissions are another byproduct of manufacturing and are a serious concern of the global society. Industrial production is a heavy consumer of fossil fuels through direct on-site combustion and indirect utilization of fossil-fuel-generated electricity—all of which contributes to the generation of significant amounts of greenhouse gases globally. Greenhouse gases mainly induce global warming problems and may cause dangerous anthropogenic interference with the climate system [43]. According to the Intergovernmental Panel on Climate Change (IPCC), based on current emission trends the average global temperature is expected to rise by 1.4–5.8°C between 1990 and 2100 [44].

In order to stabilize the concentrations of greenhouse gases in the atmosphere to prevent climate change, the Kyoto protocol, an international agreement, was reached in 1997 for establishing legally binding targets for reduction of four greenhouse gases including carbon dioxide, methane, nitrous oxide, sulfur hexafluoride, and two groups of greenhouse gases including hydrofluorocarbons and perfluorocarbons [44]. By the end of 2008, the Kyoto protocol had been signed and ratified by a total of 183 countries and regions [45]. Based on the protocol, industrialized countries will cut down their greenhouse gas emissions by a target percentage value on the base year of 1990. The initial targets were set at 8% reductions for European Union nations, 7% for the United State, and 6% each for Canada, Japan, Poland, and Hungary [46].

As different greenhouse gas emissions have different global warming effects when released into the atmosphere, the Kyoto protocol adopts the Global Warming Potential (GWP) as a standard metric for the global warming effect characterization. GWP is a standard metric developed by IPCC for the trade-off between emissions of different greenhouse gases. The GWP index is defined as the ratio of the time-integrated global mean Radiative Forcing (RF) of a pulse emission of 1 kg of a gas (i) relative to that of 1 kg of a reference gas, as expressed in the following formula (1.1) [47].

$$
GWP_i = \frac{\int_0^{TH} RF_i(t)dt}{\int_0^{TH} RF_r(t)dt} = \frac{\int_0^{TH} \alpha_i \cdot [C_i(t)]dt}{\int_0^{TH} \alpha_r \cdot [C_r(t)]dt} \tag{1.1}
$$

where

TH: time horizon;

RF_i: the global mean radiative forcing of gas i;

α_i: the RF per unit mass increase in atmospheric abundance of gas i (radiative efficiency);

$(C_i(t))$: the time-dependent abundance of gas i.

In the GWP metric, typical time horizons used for calculations are 20, 100 and 500 years, while the 100 year time horizon is the most commonly used in various analyses and statements. For example, the Kyoto protocol uses the GWP results calculated from a 100 year time horizon. In GWP calculations, the reference gas is commonly selected as CO_2 on which the GWP is set as 1. In (1.1), the numerator and denominator are called the absolute global warming potential (AGWP) of gas i and r, respectively.

As the GWP sets a cut-off time horizon for the global warming effects which are different from the life time of the greenhouse gas i, the adequacy and applicability of the GWP metric has been widely debated since its introduction [48, 49]. The 100 year time horizon is used in the Kyoto Protocol, and it has been noted that the effect of current emissions reductions that contain a significant fraction of short-lived gases (for example, CH_4) will give less temperature reductions towards the end of the time horizon, when compared to the effects of reductions in CO_2 emissions only [50]. In such a manner, the Global Warming Potentials can really be expected only to produce identical changes in one measure of climate change— integrated temperature change following emissions impulses and only under a particular set of assumptions [49, 50].

Based on (1.1), the global warming effects of various greenhouse gases can be quantified on the same scale and benchmarked for emission control and environmental management. Table 1.1 lists the GWP values of the three most common greenhouse gases [50, 51].

Through the global warming potential values, different greenhouse gas emissions can be benchmarked both on an individual basis and on a combined basis. For example, taking the 100 year time horizon, methane has a global warming effect 25 times that of carbon dioxide, following the GWP results in Table 1.1. Accordingly, for a group emission of 1 kg CH_4 and 1 kg CO_2, the combined global warming effect would be equivalent to 26 kg CO_2, which is the typical way the

Table 1.1 Global warming potential of typical greenhouse gases

Greenhouse gas	Lifetime (years)	Radiative efficiency (W/m²/ppb)	Global warming potential for given time horizon		
			20 years	100 years	500 years
CO_2	120	1.4×10^{-5}	1	1	1
CH_4	12	3.7×10^{-4}	72	25	7.6
N_2O	114	3.03×10^{-3}	289	298	153

Fig. 1.7 U.S. carbon emissions of economic sectors with electricity-related emissions distributed

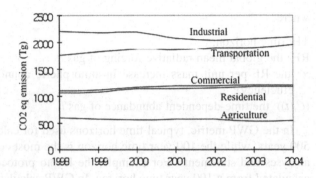

greenhouse gas emissions are evaluated and expressed for their total global warming effect in a sustainability assessment.

The industrial emissions of greenhouse gases are very significant in terms of the GWP characterized CO_2 equivalent amount. Figure 1.7 shows the greenhouse gas emissions of U.S. economic sectors between 1998 and 2004 [52]. Based on the results, the industrial sector has the largest carbon emissions among all economic sectors, while manufacturing industry typically accounts for 80% of industrial CO_2 emissions [53]. As can be seen in Fig. 1.7, the total carbon emissions from the manufacturing industry were even larger than that from transportation before 1998. But the gap is decreasing as the vehicles used for transportation undergo a continuous growth while the U.S. industry is putting great efforts to reduce its carbon emissions through either energy efficient programs or outsourcing to other regions.

The carbon emissions from the manufacturing industry result mainly from the fossil fuel energies which are consumed both directly and indirectly in manufacturing. Following the earlier discussion about automobile manufacturing and its impacts, it is helpful to consider another aspect of automotive manufacturing. Manufacturing a "typical" car requires about 120 GJ of energy input, which would generate about 23 metric tons of CO_2 and many other pollutants during the manufacturing and upstream material production processes from such an energy input [54].

Manufacturing consumes a lot of energy and materials during the processes of making products and the carbon emissions and various environmental pollutants are

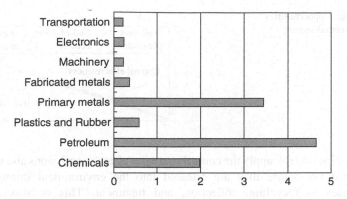

Fig. 1.8   Carbon emission of manufacturing (ton CO_2 eq./$10,000)

all embedded into the final products of manufacturing. One may not see them but they are there! Figure 1.8 shows the carbon emissions of eight typical manufacturing industries in the amount of metric tons per 10,000 dollar value produced [34]. The results show that the petroleum industry has the largest CO_2 emissions per $10,000 value generated, followed by primary metals and chemicals, while the manufacturing of consumer products such as electronics, transportation, and machinery have much lower carbon emissions, when compared with petroleum, primary metals, and chemicals such material manufacturing industries.

1.4   Strategies for Green Manufacturing

There will be more detailed discussions about strategies for greening manufacturing in the rest of this book. These discussions will go well beyond considering only environmental impacts. First a broader view of some strategies will be taken.

For industrial practice, the fundamental strategy for environmental impact reduction is environmental emission control and impact remediation. Manufacturing emissions are generated from a wide range of manufacturing activities and could be generally categorized into three groups: air emissions, water discharges, and solid wastes. Controlling such emissions requires different strategies and techniques. Generally, there are three opportunities for environmental control of emissions and wastes: pollution prevention, end-of-pipe control, and environmental restoration, as shown in Fig. 1.9 [55].

1. Pollution prevention: apply the emission control strategies before and during the emission generating process through such preventive measures as using less materials and energy and employing environmentally benign materials. This is an effective "retrofitting" technique when existing equipment or processes need to be improved. It is more effective than (2) below since it reduces the need for treatment or process waste.

Fig. 1.9 Three opportunities for environmental impact control

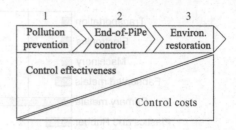

2. End-of-pipe control: apply the control strategies after the emissions and waste all generated but before they are released into the environment through such techniques as recycling, collection, and treatment. This is also a prime "retrofitting" technique for existing equipment or processes.
3. Environmental restoration: this is the environmental strategy typically employed to remediate environmental damage after the emissions/waste have been generated and released into the environment. Current environmental restoration strategies are mainly applied on land releases for hazardous waste management and site restoration, some on water treatment and just a few on airborne emission management. Environmental restoration is costly when compared with the other two strategies. This is the "oops" strategy and is usually due to a process or system out of control.

As illustrated in Fig. 1.9, the environmental impact control costs increase but the control effectiveness decreases as the strategic approach moves from pollution prevention to environmental restoration. From the perspectives of both controlling costs and controlling effectiveness, pollution prevention is the best strategy among these three environmental control opportunities for reducing the environmental impact of manufacturing [55, 56].

With a focus on the process stages prior to manufacturing, the pollution prevention strategy could enable the manufacturing system to be designed and optimized with a minimum amount of waste and emissions output. In fact, this strategy will allow the determination, and reduction or elimination of substantial process or product related impacts and resource consumption before manufacturing. This is the preferred approach, and most cost effective and will be described in more detail in following chapters.

References

1. Hawken P, Lovins A, Lovins L (1999) Natural capitalism: creating the next industrial revolution. Little Brown and Co, New York
2. New York Times (2009) House passes bill to address threat of climate change. http://www.nytimes.com/2009/06/27/us/politics/27climate.html. Accessed 1 Sep 2010
3. Department of Energy (DOE) (2010) Electric power industry 2008: year in review. http://www.eia.doe.gov/cneaf/electricity/epa/epa_sum.html. Accessed 1 Sep 2010
4. Get Energy Active (2010) Keep our fuel mix diverse. http://www.getenergyactive.org/fuel/state.htm or http://www.getenergyactive.org/fuel/mix.htm. Accessed 1 Sep 2010

5. Ford (2010) Ford ECOnetic Cars. http://www.ford.co.uk/Cars/FordECOnetic. Accessed 1 Sep 2010
6. Merriam Webster Dictionarty (2010) Define: green. http://www.merriam-webster.com/dictionary/green; accessed July 8, 2010.
7. Department of Commerce (DOC) (2007) How does commerce define sustainable manufacturing? http://www.trade.gov/competitiveness/sustainablemanufacturing/how_doc_defines_SM.asp. Accessed 12 July 2010.
8. Mas (2010) 7 Wastes. http://www.swmas.co.uk/transition/index.php/7-Wastes. Accessed 12 July 2010
9. Biles WE (2007) Environmentally benign manufacturing. In: Kutz M (ed) Environmentally conscious manufacturing. Wiley, Hoboken
10. World Commission on Environment and Development Report (WCED) (1987) chaired by Gro Harlem Bruntland. http://www.un-documents.net/wced-ocf.htm. Accessed 12 July 2010
11. Graedel T, Howard-Grenville J (2005) Greening the industrial facility. Springer, New York, 126
12. Hart SL (1995) A natural resource-based view of the firm. Acad Manage Rev 20(4):986–1014
13. Porter ME, Linde CVD (1995) Green and competitive: ending the stalemate. Harv Bus Rev 73(5):120–134
14. Ahmed NU, Montagno RV, Firenze RJ (1998) Organizational performance and environmental consciousness: an empirical study. Manage Decis 36(2):57–62
15. Klassen RD, Whybark DC (1999) The impact of environmental technologies on manufacturing performance. Acad Manage J 42(6):599–615
16. Lloyd WF (1883) Two lectures on the checks to population. Oxford Univ Press, Oxford, England. Reprinted in part in Population, evolution, and birth control (1964) In: Hardin G (ed) Freeman, San Francisco
17. Frosch RA (1995) Industrial ecology: adapting technology for a sustainable world. Environ Mag 37(10):16–37
18. Gungor A, Gupta SM (1999) Issues in environmentally conscious manufacturing and product recovery: a survey. Comput Ind Eng 36:811–853
19. US Department of Energy (US DOE) (2007) Enhancing DOE's voluntary reporting of greenhouse gases (1605(b)) program. http://www.pi.energy.gov/enhancingGHGregistry/. Accessed 15 Jan 2009
20. Organization for Economic Co-operation and Development (2007) Workshop on sustainable manufacturing production and competitiveness. Copenhagen, Denmark, June 21–22
21. Organization for Economic Co-operation and Development (2008) International conference on sustainable manufacturing. rochester, New York, USA, September 23–24
22. Bansal P (2005) Evolving sustainability: a longitudinal study of corporate sustainable development. Strategic Manage J 26:97–218
23. Rusinko CA (2007) Green manufacturing: an evaluation of environmentally sustainable manufacturing practices and their impact on competitive outcomes. IEEE Trans Eng Manage 54(3):445–454
24. Thurston D, Bras B (2001) Systems level issues: WTEC panel report on environmentally benign manufacturing. http://www.wtec.org/loyola/ebm/ Accessed 16 Jan 2009
25. Delmas M (2001) Stakeholders and competitive advantage: the case of ISO 14001. Prod Oper Manage 10(3):343–358
26. Miller WH (1998) Cracks in the green wall. Ind Week 247(2):58–65
27. Johnstone N (2007) Environemntal management, performance and innovation: evidence from OECD manufacturing facilities. OECD workshop on sustainable manufacturing and competitiveness, Copenhagen, Denmark, June 21–22
28. Seidel RHA, Shahbazpour M, Seidel MC (2007) "Establishing Sustainable Manufacturing Practices in SMEs". Proceedings of the second international conference on sustainability engineering and science, Auckland, New Zealand, February 20–23

29. Shahbazpour R, Seidel R (2006) "Using sustainability for competitive advantage". Proceedings of the 13th CIRP international conference on life cycle engineering, Leuven, Belgium
30. Freeman PK (1994) Integrating environmental risk into corporate strategy. Risk Manage 41 (7):54–59
31. Judge WQ Jr, Krishnan H (1994) An empirical investigation of the scope of firm's enterprise strategy. Bus Soc 33(2):167–191
32. Kim BR, Kalis EM, Adams JA (2001) Integrated emissions management for automotive painting operations. Pure Appl Chem 73(8):1277–1280
33. Hunkeler D, Rebitzer G (2005) The future of life cycle assessment. Int J Life Cycle Assess 10 (5):305–308
34. Gutowski TG (2004) Design and manufacturing for the environment. In: Grote KH, Antonsson EK (eds) Handbook of mechanical engineering. Springer, New York
35. U.S Environmental Protection Agency (US EPA) (2008) Toxic release inventory (TRI): chemical report. http://www.epa.gov/triexplorer/. Accessed 15 Jan 2009
36. European Environmental Agency (EEA) (2003) The European pollutant emission register. http://eper.eea.europa.eu/eper/. Accessed 18 Dec 2008
37. Organization for Economic Co-operation and Development (OECD) (2001) Why pollutant release and transfer registers (PRTRs) differ: a review of national programmes. ENV/jm/mono (2001)16. OECD environment, health, and safety publications, series on pollutant release and transfer registers no. 4.Paris: OECD
38. Park SH, Labys WC (1998) Industrial development and environmental degradation: a source book on the origins of global pollution. Edward Elgar Pub Ltd, Cheltenham
39. Wernic IK, Herman R, Govind S, Ausubel JH (1996) Materialization and dematerialization: measures and trends. Daedalus 125(3):171–198
40. Energy Information Agency (EIA) (2004) Annual energy review 2003. http://www.eia.doe.gov/emeu/aer/contents.html. Accessed 16 Dec 2008
41. Energy Information Agency (EIA) (2007) Net generation by energy source by type of producer. http://www.eia.doe.gov/cneaf/electricity/epa/epat1p1.html. Accessed 15 Feb 2008
42. Energy Star (2009) http://www.energystar.gov/. Accessed 25 Mar 2009
43. United Nations (1992) United Nations framework convention on climate change. http://unfccc.int/resource/docs/convkp/conveng.pdf. Accessed on 25 Feb 2009
44. Intergovernmental Panel on Climate Change (IPCC) (2001) Climate change 2001: the scientific basis. IPCC third assessment report. http://www.grida.no/publications/other/ipcc_tar/?src=/climate/ipcc_tar/wg1/index.htm. Accessed 8 Jan 2009
45. United Nations Framework Convention on Climate Change (UNFCCC) (2009a) Kyoto protocol: status of ratification. http://unfccc.int/files/kyoto_protocol/status_of_ratification/application/pdf/kp_ratification.pdf. Accessed 16 Jan 2009
46. United Nations Framework Convention on Climate Change (UNFCCC) (2009b) Kyoto protocol reference manual on accounting of emissions and assigned amounts. http://unfccc.int/kyoto_protocol/items/3145.php. Accessed 19 Feb 2009
47. Intergovernmental Panel on Climate Change (IPCC) (1990) Climate change: the intergovernmental panel on climate change scientific assessment In: Houghton JT, Jenkins GJ, Ephraums JJ (eds) Cambridge University Press, Cambridge, p 364
48. Fuglestvedt JS, Berntsen TK, Godal O, Sausen R, Shine KP, Skodvin T (2003) Metrics of climate change: assessing radiative forcing and emission indices. Clim Change 58:267–331
49. O'Neill BC (2000) The jury is still out on global warming potentials. Clim Change 44:427–443
50. Forster P, Ramaswamy V, Artaxo P, Berntsen T, Betts R, Fahey DW, Haywood J, Lean J, Lowe DC, Myhre G, Nganga J, Prinn R, Raga G, Schulz M, Van Dorland R (2007) Changes in atmospheric constituents and in radiative forcing. In: Solomon S, Qin D, Manning M, Chen Z, Marquis M, Averyt KB, Tignor M, Miller HL (eds) Climate change 2007: the physical science basis. contribution of working group i to the fourth assessment report of the intergovernmental panel on climate change. Cambridge University Press, Cambridge

51. Intergovernmental Panel on Climate Change (IPCC) (2005) Special report on safeguarding the ozone layer and the global climate system: issues related to hydrofluorocarbons and perfluorocarbons. http://www.ipcc.ch/pdf/session23/doc2a.pdf. Accessed 9 Jan 2009
52. US Environmental Protection Agency (US EPA) (2006) Inventory of US greenhouse gas emissions and sinks: 1990–2004. http://www.epa.gov/climatechange/emissions/downloads06/06_Complete_Report.pdf. Accessed 11 April 2008
53. Gutowski TG, Murphy CF, Allen DT, Bauer DJ, Bras B, Piwonka TS, Sheng PS, Sutherland JW, Thurston DL, Wolff EE (2001) WTEC panel report on environmentally benign manufacturing. http://www.wtec.org/loyola/ebm/. Accessed 19 Jan 2009
54. Maclean HL, Lave LB (1998) Life-cycle model of an automobile. Environ Policy Anal 3 (7):322A–330A
55. Nazaroff WW, Cohen LA (2001) Environmental engineering science. Wiley, New York
56. Zhang HC, Kuo TC, Lu HT, Huang SH (1997) Environmentally conscious design and manufacturing: a state-of-the-art survey. J Manuf Syst 16(5):352–371

The Social, Business, and Policy Environment for Green Manufacturing

2

Hazel Onsrud and Rachel Simon

> *The world will not evolve past its current state of crisis by using the same thinking that created the situation.*
>
> Albert Einstein

Abstract

The chapter introduces readers to the pressure for change, the themes of the transitions taking place, and the steps suggested for moving forward in the social, economic, and policy environment in which green manufacturing resides. The concept of sustainability related to manufacturing with an emphasis on the metrics, standards, and best practices associated with instituting green manufacturing on the path to sustainability is defined. The drivers for change and progress and the difficulties, hurdles, and benefits associated with transitioning to green operations are discussed.

Part of this chapter is work duplicated from work written by the authors as part of the UC Berkeley Sustainable Packaging Project (Towards Sustainable Packaging: Metrics, Standards & Best Practices) (http://lma.berkeley.edu/sustainablepackaging/), a project funded by the Sustainable Products and Solutions Program.

H. Onsrud (✉)
e-mail: hazel.onsrud@gmail.com

R. Simon
Laboratory for Manufacturing and Sustainability, University of California, Berkeley, 192 Highland Avenue, San Francisco, CA 94110, USA
e-mail: rachelrific@gmail.com

2.1   Introduction

Manufacturing, like farming, is one of the very few means of creating wealth. It is also a crucial component of a sustainable society. Numerous items, ranging from energy collectors to medical devices, will be required to realize a world where intergenerational and intra-generational equity is promoted and everyone has access to the items necessary to foster their well-being. In recent decades, there has been a societal shift towards sustainability, or "development that meets the needs of the present without compromising the ability of future generations to meet their own needs" [1]. Not everyone agrees on what these goals for a better future are—nor how to obtain them—but contemporary evidence suggests change is imperative. The Earth's climate is changing; the world is losing its biodiversity; and the existence of contemporary society is threatened.

2.1.1   Understanding the Need for Change: The Desire
for a Better World[1]

In contrast to the ideal systems posited by sustainability advocates and manufacturing futurists, conventional manufacturing has negative environmental and social externalities. Manufactured products and capital for production (including infrastructure, machines, tools, and factories) tend to be created though traditional structures which are inefficient and cause negative impacts. Waste, pollution (to air, land, and sea), and large ecological footprints are all signs of this inefficiency [2]. Manufacturing practices also create significant greenhouse gas emissions.[2] Yet, adapting these practices to be more efficient is difficult because the vast majority of manufacturing operations are tied to a supply chain over which individual manufacturers have little control. Many companies have been criticized for social issues associated with their supply chain, including a lack of health and safety standards, poverty prevention strategies, and measures to support any human rights abuses that occur in the communities in which they operate. Continued impacts of poverty and climate change have been pressures for manufacturers to change their practices.

---

[1] This subtitle was inspired by the book, *Alternatives to Economic Globalization: A better world is possible* [3].

[2] A World Resources Report estimates that in the year 2000 global manufacturing and construction industry emissions accounted for 21% of world's greenhouse gases [4]. Additionally, the Intergovernmental Panel on Climate Change (IPCC) 2007 report states that industry made up 19.4% of the total anthropogenic GHG emissions in 2004 in CO_2 equivalents ([5], Fig. 2.1). In the US, industrial emissions accounted for 2,510 Million Metric Tons CO_2 equivalents in 2008, a quantity larger than that year's transportation, residential, or commercial emissions [6].

Due to these numerous harmful impacts, manufacturers face significant pressure to undertake quick and comprehensive change. Such transformations require movement beyond a mere awareness of green concepts to a fast and wide-ranging implementation of better practices. Despite daunting challenges of issues to address, forerunners have illustrated that appropriately transitioning to green business practices can be very lucrative; environmentally friendly and socially responsible policies can improve a firm's public image and mitigate risks resulting from inaction.

2.1.2   Values and Practices Are Changing

Because a number of societal actors have recognized these issues, and have subsequently taken action, modern manufacturers have been caught in the midst of a number of societal, corporate, and policy transitions. Because manufacturers affect numerous social and environmental issues, stakeholders are demanding more than business-as-usual practices and policy-makers are targeting manufacturers for change, relying on the specialized knowledge and ingenuity of the manufacturer to solve their concerns. Manufacturers are being pressured to help create a "better" world, where better is interpreted differently by the various stakeholders. As a result of this pressure, corporations are being forced to change to more socially aware and environmentally conscious operations and to effectively communicate this to their stakeholders and the public.

The distinct potential for manufacturers to take a leadership position and push society towards more sustainable practices is recognized at a number of governance scales. Past experience has shown that manufacturing can revolutionize systems if placed in the right environment and with the right incentives, as it did during the industrial revolution. Hargroves and Smith [7] created a diagram noting six "Waves of innovation of the first and the next industrial revolution" illustrating that innovations facilitated by manufacturers throughout the past couple of centuries, have supported major societal transitions from "steam power" to "information technology" and that the newest wave of innovation involves ideas like "renewable energy", "green chemistry" and "radical resource productivity."

2.1.3   Concept of Sustainability

People that are concerned about the sustainability of our current lifestyles are a very vocal and influential contingent searching for a "better" world. The concept of "sustainability" has been used to convey a number of meanings; however, at its most basic, sustainability implies that we as humans want our existence and or our way of life to be "maintained." Although the concept of sustainability is ambiguous and abstract, many people are striving for this goal precisely because of its pliable definition. Although the concepts of sustainability, the plethora of its associated mechanisms, and the means of achieving it are heavily debated, it continues to be touted as a new ideal [8]. Sustainability theorists that adhere to this definition are arguing for the creation of fundamentally different societal systems based on new values. Their arguments have attracted stakeholders with various concerns to coordinate their efforts for change under the banner of "sustainable development," although these actions may be more specifically categorized under other domains such as "corporate social responsibility," "ecological design" or "industrial ecology." Because of the increasing use of the term "sustainability," a number of different conceptual frameworks have been used to focus the concept's meaning. A Venn diagram of three overlapping aspects consisting of "people, prosperity, and planet," or "equity, economy, and ecosystem" (along with a plethora of other figures) have been used to illustrate the overarching notions associated with this concept [9]. This was shown in Fig. 1.1 relative to manufacturing.

The goals and actions associated with this idea have a large conceptual range and often conflict. For example, some business people are revaluating the purpose of business, and its ability to help those in need. Instead of producing more nonessential products, some business people are targeting the needs of the larger, less-affluent proportion of the world population because they have seen the potential to grow their customer base and help people obtain with essential goods. Others critique these actions when they take money out of the country by not supporting local manufacturing operations and skills. Still other actors tackle different systemic issues with contemporary practices. For example, some people have identified issues with the current economic model and are working to change the focus of its metrics of success in order to restructure the rules under which all businesses must operate; alternative metrics to Gross Domestic Product (GDP) have been suggested; the idea of growth has been challenged; and alternatives like a steady state economy have been proposed [2, 7, 10, 11]. Overall this transition to sustainability supports a more holistic view and dialog about what society values and wants to foster, support, or persevere. Thus, in many cases, those advocating for sustainability are advocating for a fundamental restructuring of the values of our modern societal systems where a greater focus on natural limits can yield opportunities to create a "better" world. Chapter 5 will explore what goals these domains may include at the entity scale. Although, differing in their tone, motivation, and strategy of approach, many sustainability theorists are arguing that society has to restructure itself to fit within the bounds of Earth's natural systems. Different theorists and practitioners have approached this transition towards sustainability with varying foci [9]. For example, McDonough and Braungart [12], focus their goals on the transition of industrial systems, while Elkington [13] concentrates on alternative business principles. Others emphasize the role of mechanisms and tools to incite change or facilitate the transition to a system more in line with principles which govern the natural world [11].

This chapter, and the remainder of the book, will explore the role of contemporary manufacturing in the transition towards sustainability and discuss the current social, business, and political environments in which green manufacturers are operating. The remainder of the chapter is dedicated to a survey of the social, business, and political milieu surrounding green manufacturing, and the difficulties, hurdles, and benefits associated with these environments when attempting to transition to a green operations.

2.2   The Social Environment—Present Atmosphere and Challenges for Green Manufacturing

> Interdependence is and ought to be as much the ideal of man as self-sufficiency. Man is a social being.
>
> Mohandas Gandhi

This chapter will illustrate that manufacturers are operating in a modern social environment determined by numerous stakeholders, both internal and external to their company, each of whom has concerns about how the company functions. However, the ideal role of a manufacture in a world that is marked with entrenched disparities and is threatened by significant climate risks remains undefined. Conventions for addressing social concerns are developing, although when attempting to address these types of issues manufacturers have to deal with the ambiguous nature of sustainability and the wide range of its metrics. Choosing what considerations to address is not a straightforward exercise. With this in mind, this chapter segment examines some of the implementation issues involved in addressing social issues. Chapter 3 will discuss current implementation of social metrics.

2.2.1   Applying Principles of Sustainability

How to successfully proceed and reap the benefits of utilizing sustainable principles and addressing stakeholders concerns is no easy task. Once convinced to take the step towards a sustainable enterprise, businesses have to decide on their new goal, strategy, approach, rate of change, and speed of exposure. These and a host of other factors will influence their ability to succeed and to ensure their transition to a more environmental and socially responsible company is advantageous. There are a number of scales a company can choose to address and a number of concerns to tackle. McElhaney [14] defines the magnitude of the various interpretations of corporate responsibility, by differentiating the role of a business as a company, as part of the community, as part of the industry and as part of the world. Each of these roles, McElhaney [14] implies, provides different opportunities for businesses to contribute to their areas of influence. He suggests scaled recommendations ranging from "Give something back" to "Transform an industry."

A variety of internal and external frameworks exist to help facilitate the transition of individual companies and their stakeholders, as well as compare those companies to others.

2.2.2 Stakeholders

The modern environment in which manufacturers operate suggests that an implementation of both internal and external provisions will be needed to support social well-being. Social concerns at the internal level are commonly addressed by companies in terms of job security, benefits, and safety. Although, external measures that extend beyond employees working within the company's bounds have become more common. These measures address the needs of a variety of stakeholders, each with unique concerns. These stakeholders include the larger manufacturing industry as well as civil society agents (consumers, families of employees, and the community at large, which is represented by nongovernmental organizations (NGOs) and media) and governments.

2.2.2.1 Internal

A company can influence the satisfaction of internal stakeholders, through rewards, pay, and job safety and health. For this discussion, the interests of shareholders are not of great interest, as it is a company's legal obligation purpose to make profits. Meanwhile what is the appropriate level of provisions for equality, safety, and support for employees remains opaque and has different interpretations amongst various companies and their stakeholders [15]. Also up for debate is how far back in the value chain a company is responsible for working conditions. This may be more of a legal definition than an engineering one.

The treatment of employees is one of the most discussed issues regarding social sustainability. Concerns about working conditions have grown out of some highly visible problematic cases. Yet, beyond gross abuses, the precise definition of what constitutes decent working conditions is highly debatable. In general, the baseline for working standards includes the prohibition of human rights violations, such as forced labor, sweatshops, child labor, and high fatality rates. There are multiple diverging views of what entities are ultimately responsible for the well-being of individuals with which a company interacts. Some researchers are of the opinion that governments should be held accountable for ensuring the quality of life for their citizenry. Others believe it is the obligation of employers to provide a minimum standard of living for their employees. Still others view it as the individual's responsibility to create their own standards of living.

Yet manufacturers are at risk of intense criticism if any aspect of their supply chain is executed under inhumane working conditions. As with direct employment, issues of what working standards should be throughout the supply chain, and who is ultimately responsible for them arise. While manufacturers only have so much control over the working conditions of their suppliers, they do have a choice in where they source their components from, and can leverage this power to influence the behavior of their suppliers. Yet, it is unclear how deep into the supply chain a specific set of working conditions should be upheld or verified. Manufacturers generally have working relationships with their direct suppliers, which can make it easier for them to negotiate adherence to particular specifications. However, the same does not necessarily hold true beyond first tier suppliers.

All of these factors leave the most appropriate social standards for the treatment of employees ambiguous. Baseline standards exist to ensure that manufacturers are at least be cognizant of the human rights standards they should maintain. Such standards can be taken from Social Accountability International's SA 8000 standard, and the UN's International Labour Organization (ILO) conventions, Universal Declaration of Human Rights, and Convention on the Rights of the Child.

However many manufacturers do not recognize that employee standards offer an opportunity to gain strategic advantage. These standards hold some importance to consumers, as evidenced by their willingness to pay for products that are created under good working conditions [16, 17], such as fair trade products. Consumers have become increasingly critical of the working conditions under which products are created. While the public may not be aware of the exact conditions under which employees work, past incidents of poor working conditions have brought bad publicity to some of the world's largest producers.

Hawken et al. [2] offer an alternative interpretation of a company's workforce as human capital, for which any investments into generally yield a return. From this viewpoint, expenditures on the welfare or development of employees results in increased productivity or efficiency. However, determining the best social improvements that will have the greatest returns can be difficult. For instance, Azapagic and Perdan [18] note that high rates of employment improve social welfare, but at an organizational level may indicate process inefficiencies.

Whatever standards are chosen, verification of compliance for working conditions is especially problematic. For example basic commodities extracted under intolerable working conditions are indistinguishable from the same material sourced elsewhere. Wehrfritz et al. [19] note that this is the case in Malaysia, where there have been instances of rubber and palm oil being harvested under forced labor.

2.2.2.2 External

Stakeholders outside of the company are also part of the social framework of sustainability. Logan et al. [20] define corporate citizenship as the activities that make an organization accountable to its stakeholders, including employees, shareholders, consumers, suppliers, and the communities in which they are located. Similarly, some define social accountability as the total contribution that a company makes to society. For instance, Azapagic [21] states that social accountability is "related to wider responsibilities that business has to communities in which it operates and to society in general, including both present and future generations." Meanwhile the environmental interest group, Business for Social Responsibility [22], considers corporate social responsibility as a way of conducting business that meets or exceeds societies' ethical, legal, commercial, and public expectations. With these differing views of the social scope of manufacturing impacts, the following section provides an overview of the external actors that play a role in a manufacturer's social sustainability. These actors include the larger industry, consumers, NGOs, media and the community.

Consumer interests in issues of sustainability, such as human health, depletion of resources, pollution, waste creation, and climate change, have grown over the years. In the context of this chapter, the primary interest is in consumer's purchasing behavior, which can be a driver of more sustainable production. Meulenberg [23] defines sustainable consumption as a process by which social responsibility and impacts on future generations are incorporated with the needs and desires of consumers. Similarly, the ethical consumer sees a direct connection between their purchases and social issues, and expresses it through their product choices [24]. Cohen [25] points out that these consumers, a demographic known as LOHAS (Lifestyles of Health and Sustainability), spends upwards of $300 billion each year in the USA. Thus, finding ways to clearly communicate aspects of sustainability with this segment can have substantial financial incentives for producers.

The most direct social impact manufacturers have on consumers is the health and safety aspects of a product. In terms of sustainability, the question that arises is whether producers should take the initiative to improve the health impacts of their products. Some producers, such as Unilever, have done so through their dedication to improving the nutritional value of their food items. Also manufacturers would be well served by keeping mindful of the potential of contaminates in sourcing, as they are seen as responsible for any product impacts, no matter where in the supply chain contamination occurred.

One of the major goals of sustainability assessments is to trace the effects of business onto the world in which they operate. In particular, various research efforts have been dedicated to calculating the sustainability impacts that entities have on communities. These typically relate to: the equity of relationships between businesses and communities, the standards of living in communities where businesses exist, and the investments businesses make to improve such standards (e.g., philanthropic giving and development of infrastructure). Each of these issues is difficult to even assess because there is rarely a direct causational relationship between any one company's actions and the living standards of the surrounding communities. Furthermore, any improvements of these issues can only be achieved through the efforts of multiple entities such as businesses, government, and the population.

Companies often include information about their philanthropic efforts in their annual report. In addition some methodologies exist at the product level to identify goods for the impacts they have on workers and communities. For instance, fair-trade products are certified to ensure that agricultural workers are provided certain working conditions, and that international farmers receive equitable trade premiums. Meanwhile, the methodology for assessing community impacts at the process level remains unclear. Very little research has been developed on possible approaches to assess the social implications of different manufacturing processes at the time of this writing.

In general, community-based social impacts are the most subjective and hardest to determine of all social impacts. Beyond the complexities of measuring how any particular organization contributes to social conditions, assessing what these conditions are is complex. Many efforts to measure social conditions have been at the geographic level (e.g., national and city). For instance at the national level the following sustainability assessments factor in social aspects: the Genuine Progress Indicator/Sustainable Economic Welfare [26], Human Well-Being Assessment (Well-Being Index) [27], The Human Development Index [28], and Indicators of Sustainable Development [29]. Many of these assessments attempt to adjust to replace the measure of GDP to include some social measures on issues like the population's health and education. At the urban level, Shane and Graedel [30] and the UN Centre for Human Settlements [31] have proposed methodologies to assess the sustainability of cities.

The way that one company treats its various stakeholders can influence others within the industry. Manufacturers often participate in industry level coalitions or voluntary agreements in order to build support to either drive an industry forward or to maintain the status quo. Through these associations, members can more easily influence acceptable standards for others within the industry. Some industries are notoriously associated with specific social problems, or munificently thought of and creating social benefits. By becoming more sustainable, industries have the potential to improve or maintain their impacts long term. Several methodologies have been proposed to assess the sustainability of an industry [21, 32]. These assessments not only calculate the economic, social, and environmental impacts of a given industry, but also help decision makers guide national development in a more sustainable way, through policies and subsidies.

Later in this chapter the role that government plays in affecting change from manufacturers through their policy is discussed. In addition, nongovernmental organizations push producers to improve their sustainability by working with them or fighting against them. Labor groups, such as unions, USAS, UN ILO, and Social Accountability International are well established organizations that have been challenging businesses to improve working conditions for a long time. Social rights advocacy groups such as the UN, MADRE, CorpWatch, and Global Exchange also work towards improving the impacts that manufacturers have on their internal and external stakeholders. More recently, environmental groups have been advocating for changes in production to address a multitude of issues. Media is also an important contributor to the proliferation of sustainability, as it is a means to expose consumers and the public at large to relevant issues. It can also be a gauge of the importance of any issue, because to be successful it must resonate with the public. These entities can expose abuses to employees or the environment, and the problematic nature of a product or certain types of products. However, with a growing number of alternative media sources, it is more difficult than ever for manufacturers to control their image.

2.2.3 Moving Forward

The social environment in which manufacturers operate is ambiguous, shifting, and volatile. Yet, the concerns of indecisive stakeholders can impact the image and operations of a manufacturer. Thus, an understanding of stakeholder concerns, as well as influential concepts such as sustainability, and their associated metrics are all crucial elements for helping to insulate a company from future risk. Integrating that understanding into company operations can be done at a variety of levels; perhaps most important is demonstrating an understanding of the issues and what is at stake, the intent to do something about it, and a willingness to accept suggestions. This can be done in a variety of methods: by dedicating company resources to communicating about public concerns, by accepting the help of an independent NGO consultant, or by becoming involved in progressive policy creation.

> Regardless of the stakes, at this nascent state of change, it is difficult to fault those that seek advice, and either implement it with less than optimal results or reject it based on valid counterarguments. A new business operating ethic is emerging, where many manufacturers and consumers have begun to recognize that in order to make a sustainable society everyone will have to look inward, and consider if they have truly attempted to minimize their hypocrisy. Under this new raison d'être, an individual's adherence to these intentions can be tested with the "red face test," that is, can they be confronted with questions and scrutiny without turning red from naivety or lack of effort.[3]

2.3 The Business Environment: Present Atmosphere and Challenges

> Industry is a key partner for sustainable development. We rely on industry, not only for reducing the environmental impacts of the products and services it provides us with, we also increasingly depend upon industry for the innovative and entrepreneurial skills that are needed to help meet sustainability challenges.
>
> Former UNEP Executive Director Klaus Toepfer [5]

Understanding the varied desires of their stakeholders and the new focus on sustainability, a number of businesses have begun to reexamine their goals and practices to adapt them to better meet contemporary and future needs. Some have focused on expanding their markets to include the populations of developing countries by creating desirable and durable goods for the "bottom billion," while others have focused their redesign actions on internal operations and processes [34, 35]. An international coalition of 180 companies [36] states:

> We believe that the leading global companies of 2020 will be those that provide goods and services and reach new customers in ways that address the world's major challenges—including poverty, climate change, resource depletion, globalization, and demographic shifts (p. 4).

While certain types of these efforts are sometimes critiqued by anti-globalization activists and those who would prefer initiatives focused more on local self-sufficiency, a wide variety of social entrepreneurship activities have been touted under the ever-

---

[3] This comment was prompted by a conversation with Tony Kingsbury, SPS, Haas School of Business at Berkeley, where he discussed the usefulness of "a red face test."

expanding scope of corporate social responsibility [37]. These issues extend beyond social causes to environmental concerns, and addressing these matters has caused companies to reexamine their priorities and opportunities for profit.

2.3.1   Components of the Next Transition

In attempting to address these issues manufacturers have made use of sustainability principles. Increased awareness of social and environmental concerns has incited action in business, societal, and legal domains and the influence of sustainability principles can be seen in manufacturers' attempts to address these concerns. This can best be seen by three major manufacturing trends: the design of circular product systems to replace linear product systems; the provision of services to improve the design of product systems; and the construction of information rich management systems to supplement existing information pathways.

2.3.1.1   Environmental Pressures Suggest a Transition from Linear to Circular Systems of Production

Recent decades have been marked by stresses on ecosystems, an increase in resource scarcity, growth in waste, and a change of the Earth's climate. Because manufacturing systems contribute to these issues, theorists who care about manufacturing's impact on the environment have explicated the need to convert the linear systems of manufacturing to circular systems; researchers have suggested the need for more demonstration projects to further these theories; and governments have implemented zero waste policies, attempting to scale action.

Illustrative practitioners who call for a need of a more circular system are the authors McDonough and Braungart [12]. Their concept of "cradle to cradle" extends the responsibilities of manufacturers to all phases of a product's life cycle. To design a system of industry drastically different from that encased in our most recent industrial revolution, McDonough and Braungart [12] and Hawken et al. [2] suggest that the wasteful ineffectiveness of the current industrial system could be optimized if it was redesigned to have few harmful inputs or outputs and used resources judiciously. Natural systems have been used as inspiration for these redesigns. Benyus [38] advocates integrating nature as a "model and mentor."

In an ideal "cradle to cradle" cycle products and industrial byproducts are harmlessly reintegrated into the natural ecosystem or act as food for the next industrial process. McDonough and Braungart [12] have termed these systems as ones where "waste = food." Accomplishing this feat would require production facilities to be appropriately aligned with other participants within industrial parks or cooperative industrial networks [2, 12]. Hawken et al. [2] and those interested in a larger concept of waste also suggest that our industrial systems could be redesigned to concentrate on better allocating our human resources, in other words, not wasting lives. Hawken et al. [2] suggest that increased jobs can also be brought about by a closed loop system, because 75% of labor is associated with the production phase.

Regardless of ecological inspiration, companies have shown that what they previously considered trash can have value. The minimization of waste from processes, or reduction of the raw materials needed for manufacturing—through recycling, design for the environment, disassembly, or provision of increased maintenance—are strategies that can all have positive impacts on a company's bottom line. These actions are associated with a general shift towards including traditional externalities into decision making at all levels of production. Momentum for continued, more systemic action has been building: the UN has established initiatives on sustainable consumption and production; conventional corporations like Walmart have adopted a zero waste policy; and "green" pledges or actions, have become vogue.

Despite this inspiring vision and the (albeit sparse) examples of successful synergies, there are many barriers to the design of, transition to, and maintenance of a successful circular system. Change requires effort, knowledge of alternatives, and difficult decisions. The lack of control manufacturers have over their supply chain continues to impede these possibilities; at present it is difficult to coordinate, much less reimagine, the network of these upstream and downstream entities.

2.3.1.2 The Need for Greater Influence or Control Over the Company's Resources Suggests a Transition from Product Production to Service Provision

A different approach to managing one's product is needed due to the supply chain limitations manufacturer's face. Either a symbiotic network of different companies has to be engineered so that the value of the product and its waste are increased, or individual companies can transition to a service-oriented business model, rather than simply a product-based one. Ayers [39] enumerates three existing models for inter-firm cooperation, but it seems likely that more will be needed to make substantial changes.

Certain companies have illustrated that providing a continuous service, rather than solely a product, allows them to have better control over the lifecycle of their products because they must consider the value in taking back and reusing materials. Interface, Inc. is a company that has successfully refocused its vision on the service it provides, rather than the product. After Interface's founder and Chairman Ray Anderson provided a top-down mandate that inspired and incited bottom-up action, Interface began to rapidly implement changes. While the company still provided products, modular carpeting, they are designed with a service perspective [40]. For example, rather than requiring the customer to replace an entire carpet if they ruin a small portion of it, Interface's modular carpeting design created a system where small portions can easily be replaced. This strategy has worked to both improve sales and decrease costs.

Reexamining traditional product-oriented business practices requires a different perspective amongst all of the various stakeholders on the value proposition that companies provide for customers. For example, the ideals of industrialized societies stress the importance of ownership of "things." Companies, like Interface, that have successfully challenged this concept and have discovered that consumers can be satisfied in other ways while still meeting their needs. Gertsakis, Morelli, and Ryan

[41] cite a few examples of companies who add value to their products by implementing services that support a more circular industrial system. Xerox, Herman Miller Inc., and Appliance Recycling Centers of America Inc. (ARCA) focus on fostering the "value of utilization" via company initiated take-back programs, support of secondary industries, and models of product creation and maintenance that focus on the service that the product provides [41].

The refocusing of one's operation on the service value of a product is gaining more attention as people concentrate on ways to simultaneously improve the economic, ecological, and social requirements of sustainable development. The numerous benefits of green jobs, or "blue-collar employment that has been upgraded to better respect the environment and family-supporting, career-track, vocational, or trade-level employment in environmentally friendly field" [42] have been touted as a potential component of this transition. In addition to helping the environment, people suggest that these green careers also improve domestic employment rates because they are difficult to outsource; the installation and maintenance of solar panels, insulation, or other infrastructural retrofits, are all jobs which rely on manufactured goods and must be done on location [41].

The appropriate management of the physical flows of materials is important for enabling a transition to a more sustainable state. Closed loop systems of industrial symbiosis, eco-industrial parks, and industrial networks are some of the solutions that have been proposed to help our society reach this goal. However, enabling these physical systems at a number of scales will require novel management strategies.

2.3.1.3 The Need for Greater Transparency and Control Over Operations Suggests a Transition to a Highly Integrated, Information-Rich Communication System

Entities attempting to reimagine their practices are faced with intricate supply chains and complex internal and external governance requirements. Information rich management environments offer the opportunity to better achieve the management of sophisticated coordination feats and a new level of basic transparency to improve a company's accountability. Gersakis et al. [41] suggest that managing information about production processes and controlling the distribution of information to others along the supply chain are key requirements for existing models of closed loop systems to be successful. Additionally, they note that that the creation of symbiotic networks will require two forms of integration: "vertical," between companies in different, but related, supply chain sectors, and "horizontal," between similar companies that cooperate to gain a spatial or scale advantage [41]. Transparent operations help improve accountability and better illuminate the actions of the company to those who can react accordingly. Such information sharing could also facilitate cooperation and innovation among industries. Regardless of the strategy used, numerous theorists have lauded the implications of communication between internal and external company stakeholders. For example, Allenby [43] and other industrial ecologists encourage cooperation between manufacturers and their supply chain, even though he notes that a number of new legal challenges will arise out of the massive changes taking place. For example, certain types of such attempts could put actors at risk of violating laws regarding the formation of trusts [43]. Yet, information intensive strategies could be used to support these logistical feats.

However, there is a lack of implemented tools that extend beyond company bounds to coordinate with other stakeholders striving for sustainable development. Implementing information intensive operations will require dealing with siloed processes, production, and information that have never before been made transparent. Although Strebel and Posch [44] note that there are lessons to be learned about sustainable resource management from existing recycling networks, existing tools, such as Environmental Management Systems (EMSs), may not be sufficient to facilitate this transition to an information rich environment. Additionally, scaling any changes will be challenging. Given the uniqueness of each industrial ecosystem, resulting from its specific actors, inherent values, chosen criteria, and interaction with other systems, the support devices that have been found to be successful for one industrial ecosystem cannot necessarily be easily replicated, nor appropriately adapted. Standardization of tools, processing languages, or the use of open standards may help to encourage innovative monitoring.

2.3.2 Moving Forward

Certain companies are discovering their ability to benefit from practices that address appropriate social, environmental, and economic concerns. People have realized that manufacturers, specifically, have the potential to be true leaders and innovative problem solvers, responsibly building products that meet world needs. Indeed, the past industrial revolution shows us that manufacturing can revolutionize systems if placed in the right environment with appropriate incentives. The benefits of implementing socially responsible practices and environmental standards have been extensively noted for both internal and external stakeholders. For example, a progressive and responsible company stance may improve the company's internal and external reputation, helping to retain employees and please consumers, translating into industry advantages and greater profits. Decreasing wasteful practices can also provide opportunities to save money, while improving a company's environmental performance. Additionally, actions to preempt legislation have allowed green businesses to mitigate risk and reduce insurance costs. In sum, business which have begun to think of the meeting of environmental, social, and economic objectives as opportunities, rather than constraints, have reaped rewards [2, 45, 46]. Yet, despite these potential benefits, not all industries are transitioning. The United Nations has formed broad and negative conclusions about industry progression towards sustainable development. Their conclusions note insufficient progress, hindered by a lack of partnerships, dialog, commitments and aid and the need for changes and action at all scales [33, 47].

Thus, there is a lot for manufacturers to do and a lot for them to gain. Whether attacking our systems of "stuff," "shelter," "cities," "community," "business," "politics" or "planet," there is a lot to be done and a lot of ideas [48]. Indeed, the list of required changes is so extensive that it seems prudent to remain skeptical. Not everyone is going to be on board in what Elkington [13] calls the future "Chrysalis Economy." Although many companies have begun to transition their operations, Elkington [13] posits that the future of the corporate world will be formed by four very different types of companies, each requiring very different policy mechanisms to support or deter their tendencies. He predicts industry associations, government bodies, and public entities will all have a role to play in influencing their future paths and that the use of policy to facilitate the actions of forerunners and push latecomers will be necessary [13].

2.4 The Policy Environment—Present Atmosphere and Challenges for Green Manufacturing

> We now live in interesting times. ... Clearly government, the private sector, and civil
> society all have power and responsibility here. In this tripartite world, they all therefore
> have a part to play in solving the challenges of sustainable development.
>
> Hunter Lovins [7]

Because of its global reach, extensive product variety and large number of stakeholders, the manufacturing industry is seemingly caught in a mire of regulatory actions. These actions, or even the threat of such actions influence the way manufacturers do business, and while most are meant to facilitate a transition to a better state (whether promoting transparency mechanisms or attempting to level the competitive playing field socially, economically, or environmentally) their targeted actions can often have unintended consequences. It is no surprise, therefore, that stakeholders often see the role of policy as a carrot (reward) or a stick (punishment) [49]. Some argue that the need for regulation meant their business argument for green has failed [50]; others argue that, when properly implemented, it is tool to encourage change and make sure the social and environmental externalities are included in society's actions. Because of these varied ideas, a number of policy analysts, organize the wide realm of policy in a variety of ways. For example, Beder [51] condenses these influences on environmental and social policy into to six principles: "the sustainability principle," "the polluter pays principle," "the precautionary principle," "the equity principle," "human rights principles," and "the participation principle," but other academics parse up the domain differently. Yet, regardless of one's view, a better knowledge of the policy realm may help companies mitigate risk.

2.4.1 Changing Policy Trends

As the problems being tackled by stakeholders become more complex, the policies used to address these issues has become more reliant on the ingenuity of the businesses targeted. Some companies, understanding their need to mitigate risk, also seem to be more aware and open to mitigative or precompliance action.

This practice represents a change from past strategies. Traditionally, environmental legislation has mandated specific targets involving infrastructure, chemicals, or emissions [52, 53]. However, it is often the case that modern problems faced by manufacturers can no longer simply be fixed with specific and uniform solutions, but instead requires creative and customized redesigns. For example, instead of utilizing a straightforward technological solution to one specific issue, companies are now looking at redesigning their overall processes to avoid those negative impacts. A redesign of operations could result in byproducts that would not be harmful to the environment or could be used in different production cycles. Thus, as entities have begun to tackle the more holistic redesign issues associated with sustainability, environmental policy affecting stakeholders has also shifted, emphasizing the need to solve these issues through cooperation between businesses and those concerned about the businesses' impacts (stakeholders, regulators, etc.). There now exist multiple actors which rely on cooperation to set realistic industry goals. Of course, command and control strategies and regulation are still required, but in recent years the forces of some environmental policies have also relied on the use of regulations focused on specific goals and market mechanisms to incentivize companies to take action prior to formal legal mandates [43, 52, 54–56]. More specifically, since early 1990s, policies have begun to focus more on products because of their increasing presence in the waste stream and the impact of certain products' use stages [53, 57–59].

2.4.2 Fostering Cooperation

As the number of stakeholder concerns has multiplied, and the issues involved require more holistic design solutions, certain actors have begun to understand the need to work together to create solutions to existing problems. This idea of different actor roles and the idea of governance, rather than a focus on government as a lone actor responsible for environmental policy, has become more recognized as more stakeholders flex their influence [43, 59]. Indeed, many have recognized the need for government, NGOs, academia, the public, and corporations to cooperate in order to effectively "race" towards sustainable development [60].

Each of these actors can take a number of actions to support environmental action. Government can make and enforce legal agreements, support a market environment favorable to green innovation, and fund investments necessary to facilitate large-scale industrial changes (e.g., education, infrastructure, and research). NGOs and academics can help to inform the direction of the transition and provide other actors with guidelines and standards for consumption and action. The public can be responsible and educated consumers and buy items and services that match their values; their uses of dollars are votes for or against toxicity, human rights, and habitat preservation. Companies can also take a proactive stance to make sure they help to define the industrial transitions others desire to take place; indeed, they have the great power to both clean up their operations and create items that matter and influence their overall industry [7, 57].

This is not to say that everyone will cooperate and strive for these goals willingly—it is likely that certain existing power dynamics will require significant pressure and the maintenance of a contemporary regulatory system to change; however, a totally confrontational stance by all parties may not always be necessary. All of these actors have roles to play in defining and making the transition to a green economy [7, 55, 57]. Additionally, communication between internal and external company stakeholders will be crucial to aiding this transition. For this to occur, access to the quality information, at appropriate times, and in effective formats will be crucial for the success of entities exchanging information at multiple scales and new levels of complexity [61].

2.4.3 Moving Forward

Because of the large number of stakeholders, the policy environment in which the manufacturing industry operates is both challenging to understand and maneuver through. This suggests that it is important for manufacturers to understand the potential desires of their consumers and the broader frameworks emerging in certain critiques. For example, many criticisms of action are associated with the idea of sustainability and its common interpretations (e.g., social, environmental, and economic pillars). These understandings can be integrated into a company's operations through priorities and action in order to both mitigate risk and make a difference to those goals.

Taking action can involve a number of strategies. At the most basic level, education on priority issues and a public acknowledgment of what a company can improve is a crucial step to mitigate risk and to fostering an intelligent relationship with one's buyers. These types of open and well-meaning practices, if appropriately planned, can help to both ease tensions with potential opponents and illuminate a company's priorities for a larger audience. It can also help a company find innovative solutions to their most challenging problems. Promoting a culture of sincerity, honesty, and transparency is crucial to foster cooperation rather than scathing critiques from a potential adversary.

Because of their specialized expertise, it makes sense that manufacturers will want a say in creating policies to facilitate changes in their industries. A number of ways exist for manufacturers to influence policies. They can utilize tools to facilitate dialog on policies, or they can implement voluntary action to stave off formal policy. There are often opportunities for interested stakeholders to participate in calls for industry participation. Indeed, certain organizations are required to solicit industry or general stakeholder opinions [43, 57, 62]. Although informational tools are helpful for keeping up to date on the most pertinent issues and ensuring one's company conforms to the status quo of upcoming regulatory action, as a company and as a person, it seems more likely that one would be approached by policy makers and other stakeholders, if it is apparent that one is interested in these issues and is willing to work with others who may have different views. A variety of resources exist to guide interested companies in the general direction towards a sustainable entity, but reaching the majority of benchmarks will require a significant amount of education and dedication.

Despite the critiques made of manufacturers, some believe that society simply cannot move at the rate of change required without the ingenuity, specialized knowledge, and resources of business. One can wear as many tee shirts or bumper stickers stating, "'You must be the change you wish to see in the world'—Mahatma Gandhi or 'Never doubt that a small, group of thoughtful, committed citizens can change the world. Indeed, it is the only thing that ever has.'—Margaret Mead," but it is the manufacturers of those tee shirts, bumper stickers, laundry machines, cleaning products, and cars that have the greater ability to make such change happen.

2.5   Conclusion

> They must often change, who would be constant in happiness or wisdom.
>
> Confucius

Perhaps the question is not who wants to be a green manufacturer, but rather who does not. If being part of the next industrial revolution implies greater profits and positive impacts, while remaining static endangers one's business and our society's greater survival, it seems prudent to strive for goals of a healthy ecosystem, economy, and society [2]. Yet, the state of our world is far from this aim. Manufacturing is caught in a transition where social measures are vague and difficult to prioritize, and where the dominant business model is having its validity challenged. Not only do the frameworks to guide business forward vary in depth, breadth, and proof of application, but our system of governance is complex and difficult to navigate, much less create or uphold an appropriate level of coordination and order. Despite these inherent difficulties, perhaps the most important trait of this predicament is that it does include everyone. All of society is bound to suffer from a lack of change [63–66]. Indeed, many people already are. This means society has a greater reason than ever before to not only hope that change of the needed magnitude at the required speed is possible, but to take the needed action to make it happen. Manufacturers, more than most, are uniquely positioned to contribute to this transition and reap its potential rewards. This is recognized by business people and the public alike. Indeed, when Shellenberger and Nordhaus [67] placed the Chinese ideogram for "crisis" on the cover of their essay, *The Death of Environmentalism*, they noted on the next page that this was made up of two characters, "danger" and "opportunity."

As part of this transition, a number of people are advocating that society needs to transition from a largely detrimental system of production to one that is environmentally benign. This system also must promote a healthy workplace and economy. However, at times, these goals are in conflict with our contemporary practices and dominant systems of production. Hawken et al. [2] note: "For all their power and vitality, markets are only tools. They make a good servant but a bad master and a worse religion." (p. 261). Although businesses are situated in a changing environment, filled with critiques regarding the best course of action; there are still profitable and ethical paths down which to proceed. Indeed, what Willard [46] calls a "perfect storm" of drivers for the next sustainability wave is present here and now. Impacts of climate change and globalization are made relevant to contemporary business dealings with the pressure from internal and external company stakeholders, who, although motivated by different criteria, are all concerned with the risks associated with these pressures. Because of the wide spectrum of issues to be addressed, and the speed at which change is required, the next wave of action will have to be more inclusive. We all will have a role to play.

The large number stakeholders involved in manufacturing are demanding change and require a new level of transparency, analysis, and action. Manufacturers have to be aware of the concerns of their stakeholders in order to mitigate their future risks and help build an acceptable future for all. Yet, stakeholders have to help facilitate such action.

With the ever-increasing plethora of products entering our waste stream, some might argue that we have progressed from craft to crap production; yet, this view is unjust, especially from the viewpoint of producers. Manufacturers excel at creating products to meet demand within the constraints of a larger economic system, and modern society's demands have been diverse, short-term oriented, and plentiful. Stakeholders have a role to play in helping manufacturers meet their demands, either by voting with their dollars or advocating that the government help green solutions compete.

In the future, a number of new trends will shape the role of the manufacturers. They may have to help address the needs of those suffering from lack of income and material disparities or those requiring tools to transform their society into an eco-efficient entity. People need practical means to reach their aspirations (although what those aspirations are may be an issue) and manufacturing may be able to help provide a type of this aid. These types of requests will require increased information management requirements, and increased cooperation between corporations, the public, and the governance agencies.

An examination of the policy environment in which manufacturers operate has also illustrated that government alone is no longer in charge of governance. The modern policy environment is complex and difficult to keep up to date on, much less be involved in. Yet, working towards understanding the goals of one's stakeholders and making strives to not only conform but push the boundaries of new actions towards being a sustainable company can help turn potential critics into innovation partners. Indeed it seems as though the new environment could be characterized by cooperation among forerunners, after all, we all have something at stake.

To make our transition towards a better system, we also need input from a variety of academic disciplines as well as business leaders and governance agencies who can implement the suggested changes. We are beyond a point where one solution could fix all our problems; instead, we require action on the part of all affected participants [7, 57]. Whether they include top-down guidance or freedom to incite bottom-up action, new multichannel communication will be crucial to this forward progression. This will require horizontal and vertical means to communicate as well as cooperation between disciplines to quickly adapt good ideas to unique circumstances [41].

Antoine de Saint-Exupéry posited the following strategy:

> If you want to build a ship, don't herd people together to collect wood and don't assign them tasks and work, but rather teach them to long for the endless immensity of the sea.

Yet, longing for sustainable processes and a world where each person's dreamed version of social justice, a flourishing ecosystem and a healthy economy exists will not simply emerge from imagining this dynamic view. To reach this utopia, society needs a certain amount of utilitarianism. Despite this necessary motivation, manufacturers need actual steps on which to take action.

This chapter has attempted to introduce readers to the pressure for change, the themes of the transitions taking place, and the steps suggested for moving forward in the social, economic, and policy environment in which green manufacturing resides. Neither nuanced analysis of issues nor critical documentation of solutions can be accomplished in a sole book chapter, and, certainly, these authors do not have sufficient answers. Yet, there is something to be said for taking action, however slight, at a time when we all face significant pressure for deep and widespread change.

Sustainability is not solely what the societies of the world are looking for; when thinking about the future, most people are looking for something better. As Braungart [68] notes, a happy couple would not define their relationship as merely "sustainable" and in this same vein, those looking to improve the world at this time of crisis, are looking to capture the tenacity and creatively of people in need and truly reinvent the world. Unlike many actors, manufacturers have done this before. The industrial revolution dramatically changed society; our lifestyles, our environment and our economy and our embedded values. The next industrial paradigm shift will likely have to do no less [2].

References

1. World Commission on Environment and Development (1987) Our common future. United Nations, Oxford, New York. Retrieved from http://www.un-documents.net/wced-ocf.htm. Accessed 8 Nov 2010
2. Hawken P, Lovins A, Lovins LH (2000) Natural capitalism: creating the next industrial revolution, 1st edn. Back Bay Books, New York
3. The International Forum on Globalization (2004) Alternatives to economic globalization: a better world is possible. Berrett-Koehler Publishers, San Francisco
4. Baumert KA, Herzog T, Pershing J (2005) Navigating the numbers: greenhouse gas data and international climate policy. The World Resources Institute, Washington, DC
5. Pachauri RK, Reisinger A (Eds) (2007) Climate Change 2007: synthesis report. Contribution of Working Groups I, II and III to the Fourth Assessment Report of the Intergovernmental Panel on Climate Change. Geneva, Switzerland: IPCC. Retrieved from http://www.ipcc.ch/publications_and_data/publications_ipcc_fourth_assessment_report_synthesis_report.htm. Accessed on 4 Jan 2011

6. US Department of Energy (2009) Emissions of greenhouse gases in the United States 2008. US Department of Energy Washington, DC. Retrieved from ftp://ftp.eia.doe.gov/pub/oiaf/1605/cdrom/pdf/ggrpt/057308.pdf. Accessed 8 Nov 2010
7. Hargroves K, Smith MH (2005) The natural advantage of nations: business opportunities, innovation and governance in the 21st century (illustrated edn.). Earthscan Publications, London
8. Komiyama H, Takeuchi K (2006) Sustainability science: building a new discipline. Sustain Sci 1(1):1–6
9. Edwards AR (2005) The sustainability revolution: portrait of a paradigm shift. New Society Publishers, Gabriola Island
10. Jackson T (2009) Prosperity without growth? The transition to a sustainable economy. UK Sustainable Development Commission. Retrieved from http://www.sd-commission.org.uk/publications.php?id=914. Accessed 8 Nov 2010
11. Wheeler SM, Beatley T (2004) The sustainable urban development reader, 1st edn. Routledge, New York
12. McDonough W, Braungart M (2002) Cradle to cradle: remaking the way we make things, 1st edn. North Point, New York
13. Elkington J (2001) The chrysalis economy: how citizen CEOs and corporations can fuse values and value creation. Capstone, Oxford
14. McElhaney KA (2008) Just good business: the strategic guide to aligning corporate responsibility and brand. Berrett-Koehler Publishers, San Francisco
15. Marks N, Simms A, Thompson S, Abdallah S (2006) The (Un)happy planet index. New Economics Foundation, London
16. Hiscox MJ, Smyth NFB (2006) Is there consumer demand for improved labor standards? Evidence from field experiments in social labeling. Department of Government, Harvard University
17. Elliot KA, Freeman RB (2003) Can labor standards improve under globalization? Institute for International Economics, Washington, DC
18. Azapagic A, Perdan S (2000) Indicators of sustainable development for industry: a general framework. Trans IChemE B 78(4):243–261
19. Kinetz E, Wehrfritz G, Kent J (2008) Bottom of the Barrel. MSNBC, Newsweek, Mar 24. Retrieved from http://www.newsweek.com/2008/03/15/bottom-of-the-barrel.html. Accessed 1 Oct 2010
20. Logan D, Roy D, Regelbrugge L (1997) Global corporate citizenship—rationale and strategies. The Hitachi Foundation, Washington, DC
21. Azapagic A (2003) Systems approach to corporate sustainability: a general management framework. Trans IChemE B 81:303–316
22. Business for Social Responsibility, (1998) Introduction to corporate social responsibility. BSR, San Francisco
23. Meulenberg M (2003) Consumer and citizen, meaning for the market of agricultural products and food products. TSL 18:43–56 (In Dutch)
24. De Pelsmacker P, Driesen L, Rayp G (2003) Are fair trade labels good business? Ethics and coffee buying intentions. Working papers of faculty of economics and business administration. Ghent University, Belgium 3(165)
25. Cohen MJ (2007) Consumer credit, household financial management, and sustainable consumption. Int J Consumer Stud 31(1):57–65
26. Cobb C, Halstead T, Rowe J (1995) The genuine progress indicator: summary of data and methodology. Redefining Progress, Washington, DC
27. Prescott-Allen R (2001) The wellbeing of nations: a country-by-country index of quality of life and the environment. Island, Washington, DC
28. UN Development Programme (2005) Human Development Report: International cooperation at a crossroads: aid, trade and security in an unequal world. United Nations, New York

29. United Nations Division for Sustainable Development (DESA) (2010) Division for sustainable development and UNEP—division of technology, industry and economics. Available at http://webapps01.un.org/dsd/scp/public/Welcome.do. Accessed 30 Aug 2009
30. Shane AM, Graedel TE (2000) Urban environmental sustainability metrics: a provisional set. J Environ Plan Manage 43:643–663
31. United Nations Centre for Human Settlements (2001) The state of the worlds cities report. UN, Nairobi
32. Singh RK, Murty HR, Gupta SK, Dikshit AK (2007) Development of composite sustainability performance index for steel industry. Ecol Ind 7:565–588
33. United Nations Environment Programme (UNEP) (2006) Class of 2006: industry report cards on environment and social responsibility. UNEP, Nairobi
34. Bornstein D (2007) How to change the world, Updatedth edn. Oxford University Press, Oxford
35. Collier P (2007) The bottom billion: why the poorest countries are failing and what can be done about it, 1st edn. Oxford University Press, New York
36. World Business Council for Sustainable Development (2006) from challenge to opportunity: the role of business in tomorrow's society. Available at http://www.wbcsd.org/plugins/DocSearch/details.asp?type=DocDet&ObjectId=MTgyMTM. Accessed 30 Aug 2009
37. Hopkins M (2007) Corporate social responsibility and international development: is business the solution? Earthscan Publications Ltd., Sterling, VA
38. Benyus JM (2002) Biomimicry: innovation inspired by nature. Harper Perennial, New York
39. Ayers RU (2002) On industrial ecosystems. In: Ayers RU, Ayers LW (eds) A handbook of industrial ecology. Edward Elgar Publishing, Northampton
40. Interface (2004) Nature and the industrial enterprise: mid-course correction. Engineering Enterprise, Spring. Retrieved from http://www.interfaceglobal.com/getdoc/a7aa467e-801d-44c0-aa41-6b410ef1881e/Interface_Sustainable_Model.aspx. Accessed 1 Oct 2010
41. Gertsakis J, Morelli N, Ryan C (2002) Industrial ecology and extended producer responsibility. In: Ayers RU, Ayers LW (eds) A handbook of industrial ecology. Edward Elgar Publishing, Northhampton
42. Jones V (2008) The green collar economy: how one solution can fix our two biggest problems. Harper Collins, New York
43. Allenby BR (2002) Industrial ecology: governance, laws and regulations. In: Ayers RU, Ayers LW (eds) A handbook of industrial ecology. Edward Elgar Publishing, Northampton
44. Strebel H, Posch A (2004) Interorganisational cooperation for sustainable management in industry: on industrial recycling networks and sustainability networks. Prog Indu Ecol Int J 1(4):348–362
45. Hart SL (2007) Capitalism at the crossroads: aligning business, earth, and humanity, 2nd edn. Wharton School Publishing, Upper Saddle River, NJ
46. Willard B (2005) The next sustainability wave: building boardroom buy-in. New Society Publishers, Gabriola Island, BC
47. United Nations Environment Programme (2002) Global status 2002: sustainable consumption and cleaner production. United Nations Environment Programme, Division of Technology, Industry and Economics, Paris
48. Steffen A (2008) Worldchanging: a user's guide for the 21st century. Harry N. Abrams Inc, New York
49. Wilson DC (1996) Stick or carrot? The use of policy measures to move waste management up the hierarchy. Waste Manag Res 14(4):385–398
50. Shulman K (2001) Think Green. Metropolis Magazine, August/September. Retrieved from http://www.metropolismag.com/html/content_0801/mcd/. Accessed on 4 Jan 2011
51. Beder S (2007) Environmental principles and policies: an interdisciplinary introduction, 1st edn. Earthscan Publications Ltd., Sterling, VA
52. Anderson D (2008) Regulatory policy vs economic incentives. The Environmental Literacy Council. Retrieved from http://www.enviroliteracy.org/article.php/1329.html. Accessed 13 Aug 2009

53. Levy GM (1999) Packaging policy and the environment, 1st edn. Springer, New York
54. Bailey I (2003) New environmental policy instruments in the European Union. Ashgate Publishing, Ltd, Burlington, VT
55. Bardach E (2008) A practical guide for policy analysis: the eightfold path to more effective problem solving, 3rd edn. CQ, Washington, DC
56. Stuart R (2007) Command and control regulation. The Encyclopedia of Earth. Retrieved from http://www.eoearth.org/article/Command_and_control_regulation. Accessed 8 Nov 2010
57. Chasek PS, Brown JW, Downie DL (2006) Global environmental politics, 4th edn. Westview, Boulder, CO
58. Hwang B (2007) Unpacking the packaging problem: an international solution for the environmental impacts of packaging waste. Retrieved from http://works.bepress.com/billy_hwang/1/. Accessed 4 Jan 2011
59. Mayntz R (2006) Chapter 1: From government to governance: political steering in modern societies. In: Scheer D, Rubik F (eds) Governance of integrated product policy: in search of sustainable production and consumption. Greenleaf Publishing, Sheffield
60. Sutton P (2004) The "Race to (environmental) Sustainability"—a proposal. Available at http://www.green-innovations.asn.au/Race-to-Sustainability.htm. Accessed 17 Nov 2010
61. Joas M, Evans B, Theobald K (2005) Evaluating governance for local sustainability: online tools for self-assessment. Prog Ind Ecol Int J 2(3–4):440–452
62. Greenberg EF (2007) Guide to packaging law: a primer for packaging professionals, 2nd edn. Institute of Packaging Professionals, Herdon, VA
63. Frosch RM, Manuel P, Sadd J, Shonkoff S (2009) The climate gap: inequalities in how climate change hurts americans & how to close the gap. University of Southern California, Program for Environmental and Regional Equity. Retrieved from http://college.usc.edu/geography/ESPE/perepub.html. Accessed 4 Jan 2011
64. Richardson K, Steffen W, Schellnhuber HJ, Alcamo J, Barker T, Kammen DM, Leemans R, et al (2009) Synthesis Report: climate change: global risks, challenges & Decisions. Copenhagen: International Alliance of Research Universities. Retrieved from http://climatecongress.ku.dk/pdf/synthesisreport/. Accessed 4 Jan 2011
65. Stern N (2006) Summary of conclusions in the Stern Review: the economics of climate change. The Office of Climate Change, United Kingdom. Retrieved from http://www.hm-treasury.gov.uk/sternreview_index.htm. Accessed 4 Jan 2011
66. World Economic Forum (2008) Global risks 2008, a world economic forum report. Retrieved from http://www.weforum.org/pdf/globalrisk/report2008.pdf. Accessed 4 Jan 2011
67. Shellenberger M, Nordhaus T (2004) The death of environmentalism: global warming politics in a post-environmental world. Retrieved from http://www.thebreakthrough.org/images/Death_of_Environmentalism.pdf. Accessed 8 Nov 2010
68. Braungart M (2008) Michael Braungart: Why less bad isn't better? Vimeo video lecture. Presented at the DO lectures. Retrieved from http://vimeo.com/2362082. Accessed 10 Nov 2010

Metrics for Green Manufacturing

3

Corinne Reich-Weiser, Rachel Simon, Timo Fleschutz,
Chris Yuan, Athulan Vijayaraghavan, and Hazel Onsrud

> *Not everything that can be counted counts, and not*
> *everything that counts can be counted.*
>
> Albert Einstein

Abstract

This chapter looks at metrics for green manufacturing and sustainability. Relevant economic metrics are reviewed and for complete coverage of sustainability issues, social metrics are also surveyed. The challenges of quantitatively evaluating social concerns are illustrated by highlighting the multiple considerations that social

C. Reich-Weiser (✉) • H. Onsrud
1016 Florence Ln, Apt. 3, Menlo Park, CA 94025, USA
e-mail: corinnerw@gmail.com; hazel.onsrud@gmail.com

R. Simon
Laboratory for Manufacturing and Sustainability, University of California, Berkeley,
192 Highland Avenue, San Francisco, CA 94110, USA
e-mail: rachelrific@gmail.com

T. Fleschutz
Department of Assembly Technology and Factory Management, Institute for Machine
Tools and Factory Management (IWF), TU Berlin PTZ2, Pascalstrasse 8-9,
TU 10587, Berlin
e-mail: fleschutz@gmail.com

C. Yuan
University of Wisconsin, Milwaukee, WI, USA
e-mail: cyuan@uwm.edu

A. Vijayaraghavan
System Insights, 2560 Ninth Street, Suite 123A, Berkeley, CA 94710, USA
e-mail: athulan@systeminsights.com

D.A. Dornfeld (ed.), *Green Manufacturing: Fundamentals and Applications*,
DOI 10.1007/978-1-4419-6016-0_3, © Springer Science+Business Media New York 2013

metrics attempt to capture. The chapter then survey metrics that tie in multiple considerations, pulling together ecological, social, and economic metrics. To inform metrics development, methods for inventory and impact assessment are also reviewed. Finally, the chapter presents several approaches for metric development, which systematically build up the metric based on considerations of goal, scope, system boundary, planning horizon, and system drivers.

3.1   Introduction

Innovative strategies are needed to achieve green and sustainable process technologies and industrial systems. "Green" technologies are often understood as those capable of meeting product design requirements while minimizing environmental impact. Minimizing impacts, however, is a necessary but not a sufficient condition for a sustainability strategy; recall Fig. 1.2 in Chap. 1.

For the purposes of this book, "sustainability"' is understood as the ability of an entity to "sustain" itself into the future without impacting the capacity of other entities in the system to sustain themselves. This definition involves consideration of three main drivers: economics, society, and the environment. The first of these, economics, has traditionally been the focus of the manufacturing research community. Societal concerns have been addressed by researchers as they relate to increased profit; however additional social metrics to be considered include poverty, gender equality, nutrition, child mortality, sanitation, health, education, housing, crime, and employment [1] as discussed in Chap. 2. Aggregated indices that provide a broad value for "well-being" or "environmental sustainability" have also been developed [2]. While these social and aggregate metrics are valuable to make broad decisions, they may not allow for granular insight and decision making within the manufacturing enterprise.

A challenge in selecting metrics for sustainable manufacturing is that it is not an inherently intuitive process. Unlike economic metrics, such as unit cost or part quality, sustainability metrics are not necessarily related to the function of the part being manufactured. Additionally, a complete picture of environmental impact and sustainability requires numerous metrics. However, time and cost considerations limit the number of possible metrics that can be practically considered in a manufacturing analysis. Choosing an appropriate set of metrics is critical as this choice will impact the conclusion of the analysis. For example, Schweimer and Levin [3] conducted an environmental life cycle assessment (LCA) of automobile manufacturing and found that 81% of CO_2 emissions occur during the vehicle use phase, 88% of non-methane volatile organic carbon (VOC) emissions occur in the fuel production phase, and 83% of dust emissions occur during the vehicle manufacturing phase. Hence, the least sustainable phase of the automobile manufacturing process can be identified only based on the goal of the assessment (that is, if the goal was to minimize VOC emissions, CO_2 emissions, or dust emissions). For efficiently selecting metrics it is very important to have the utmost clarity on the goal of the environmental assessment and the aspects that are important for a specific industry or world region.

Our work is motivated by the need to provide manufacturing engineers and scientists a set of tools with which they can better design and characterize sustainable manufacturing systems. A robust set of metrics will enable the vision outlined by researchers in the field, and will help integrate the specific advances in manufacturing technology into the broader framework of sustainable production systems.

This chapter systematically discusses existing metrics within each of the three pillars of sustainability, with the goal of providing insight into current practices and a framework to begin measuring sustainability. While all three pillars are discussed, the focus here is on "environmental" sustainability as it pertains specifically to green manufacturing. As has been discussed in Chap. 1, green manufacturing is the goal of reducing environmental impacts over time. This goal is in-line with, but distinct from, the goal of sustainability. Improvements in environmental behaviors are considered green; however, the activities may not have reached a "sustainable" level, in that they could be continued indefinitely.

3.2   Overview of Currently Used Metrics

The following sections are intended to provide a baseline for understanding existing economic, social, and environmental metrics. It is not possible here to cover all possible metrics under each category; however the aim is to provide a summary and references to additional documentation in each area. Economic metrics are discussed first as they can be extrapolated to useful environmental and social measurements and indicators.

3.2.1   Financial Metrics

Economic metrics are focused on investment decisions and increasing profitability over time. More detailed information can be found in Brealey and Myers standard book for corporate finance [4]. Further metrics for the economic valuation of products or companies are given by, e.g., the global reporting initiative [5]. Additional economic metrics exist at the macro level, for the planning and evaluation of industries and governments. However, these are beyond the scope of this book, and will not be further addressed here. Details regarding some of the economic metrics for manufacturers are outlined below.

Net Present Value (NPV): Determining whether an investment in future cost savings is worthwhile requires adjusting future cash flow into a current time frame. This is called the NPV, which discounts all future cash flows of an investment with a risk-adjusted discount rate (3.1):

$$\text{NPV} = -I_0 + \sum_{t=1}^{T} \frac{X_t}{(1+i)^t},$$ (3.1)

where X_t if the Cash Flow in Period t, I is the interest rate, and $I0$ is the investment in period 0. This approach is fundamentally similar to, and will provide the same conclusions as, a uniform annual cost calculation.

If the net present value of an investment project is above zero, it is profitable. The risk-adjusted rate can either be estimated with Capital Asset Pricing Model (CAPM) or is based on the Weighted Average Cost of Capital (WACC) to handle the effects of the systematic risk in an appropriate way.

Additional methodologies are the payback period and the internal rate of return (IRR). With the payback period, the time period until an investment is paid back with its returns is calculated. The internal rate of return, also termed the rate of return (ROR), gives the annualized effective compounded return rate of an investment.

An alternative to using the expected mean of future cash flows for evaluation of the best and worst cases is the best and worst expected cash flow. The Value at Risk (VaR) or Value at Chance (VaC) gives the negative resp. positive distance to the mean within a defined time horizon and confidence level [6]. Additionally, Monte Carlo simulation can be used for sensitivity analysis; it considers random sampling of probability distribution functions in order to model hundreds or thousands of possible outcomes of selected variables. As a result the simulation provides probabilities of different outcomes.

Life Cycle Costing (LCC): A complete cost analysis needs to be performed frequently in green manufacturing for decision making since green manufacturing-related adjustments either lead to a cost-saving or may require extra cost for operations. Decisions have to be made based on the cost and payback of the activities prior to the implementation of sustainability programs.

Current standards and guidelines define LCC as the consideration of all costs, which a system causes over its lifetime. The cost can be divided into the phases of "before use," "during use," and "after use" [7]. Seen as superior to simple acquisition cost comparison, LCC has been part of military instructional framework in the United States since the 1960s.

Woodward [8] notes that "LCC seeks to optimize the cost of acquiring, owning and operating physical assets over their useful lives by attempting to identify and quantify all the significant costs involved in that life, using the present value technique." The German norm VDI 2884 provides a general procedure for purchasing, operating, and maintaining production equipment using LCC. Based on a decision for LCC the maintenance strategy and the application conditions are defined for alternative production equipment. For relevant LCC factors, costs and yields are registered and evaluated. The choice can then be decided upon based on a qualitative or a quantitative method, applying dynamic investment decision tools. An eight-step formulation for the mathematical calculation is given by Kaufmann [9].

The European SETAC working group considers LCC not only as a cost calculation method but rather also as a controlling method for the estimation of all costs associated with an artifact. They distinguish between internal and external costs. Internal costs are all of the cash flows of the actors involved within the economic system. External costs are compensations for the product's impact on the social and ecological environment. They propose to mainly work with internal or already internalized costs by taxes or subsidies [10].

Cost of Ownership: Fundamentally similar to LCC, Cost of Ownership (CoO) is a commonly used term in manufacturing in general and has applications in green manufacturing as well.

A complete economic analysis metric is employed for cost assessment and comparisons among various alternative options in green manufacturing [17, 18]. The cost components typically considered in the CoO analysis include the equipment costs, setup costs, and annual costs for operation of facilities. The equipment costs are the purchasing costs of the equipment and associated components; setup costs include those for installation, transportation, and engineer training and also the labor cost for installing the facility on the site; annual operational costs include those for electricity, space consumed on the factory floor, maintenance, and consumables. As the equipment costs and setup costs are one-time capital costs occurring at the beginning of the project, the annual operational costs reach into the future depending on the designed lifetime of the equipment for operation. In order to compare the cost of alternatives on the same benchmark, usually people have the annual operational cost projected to the present value, based on the economic discounting method. The complete CoO model is shown below:

$$CoO = \frac{\left(C_1 + C_2 + \frac{C_3}{(1+i)^t}\right) \times N}{B \times V_E},$$ (3.5)

where
Equipment cost:

$$C_1 : \text{equipment cost}, \quad \$ \text{ in first year}$$

Setup cost:

$$C_2 = I + T + \sum P \times F, \quad \$ \text{ in first year}$$ (3.6)

I: installation fee including labor cost, \$; T: transportation fee, \$; P: number of people to be trained; and F: training fee for each person, \$.
Annual operational cost:

$$C_3 = S \times R_s + U \times R_E + \sum O + H \times R_M \ \$/\text{year}$$ (3.7)

S: footprint for the equipment, feet^2; R_S: footprint cost rate, \$/feet^2/year; U: electricity consumed, KWH/year; R_E: electricity rate, \$/KWH; O: cost of various consumables, \$; H: downtime, hours; R_M: maintenance cost, \$/hour; N: number of pieces of equipment needed for the system; i: discounting rate; t: t years into the future; B: total amount of output from the system; and V_E: economic value of unit output.

3.2.2 Metrics for Ecology

The first approaches for the assessment of environmental aspects go back to the Technology Assessment (TA) in the 1960s. As a tool for short- and long-term political decision making, TA evaluates the impact of new technologies based on system analysis, simulation models, and technical prognosis methods [19].

Metrics for ecology can be generally differentiated based upon their evaluation approach and their aggregation method [19]. A synopsis of the various evaluation approaches and aggregation methods are outlined below.

Evaluation approaches:
- *Utility analysis*: The combination of qualitative and quantitative criteria to a quantitative result, weightings of each criteria are subjective to user.
- *Harm and utility analysis*: Close to cost–utility analysis, each criteria is evaluated on a scale, the sum of all evaluations is the final output.
- *Approach of critical quantities*: For each criteria a threshold is defined; the value between current criteria value and threshold is measured and compared.
- *Hierarchical approach*: Criteria are structured according to a hierarchy.
- *Verbal-argumentative approach*: Goals/aims are defined qualitatively. Verbal-argumentative is used if quantification of criteria is not possible.

Aggregation methods:
- Low level
  - Vertical aggregation: One element is aggregated over the vertical (life cycle of a product), e.g., CO_2 emissions.
  - Aggregation to one criterion: Aggregation of several elements is converted to one criteria, e.g., greenhouse gases and global warming potential.
- High level
  - Aggregation to one value: The available information is aggregated to classical media, e.g., soil, water, and air.
  - Aggregation to multiple values: Aggregation of all elements to one criterion.

3.2.3 Metrics for Society

Social metrics have already been discussed in terms of how they pertain to policy in Chap. 2. However, many companies have also implemented social metrics as the first step towards sustainability goals. Social improvements, while difficult to measure, tend to be more straightforward to implement, and realize a big bang for the buck in public perception. The difficulty then is determining how to quantify and develop metrics around social sustainability.

Research into the social effects of productive practices—leading to the development of the so-called social metrics of sustainability—has been limited [20]. The reasons for this are various. Some researchers find these metrics to be too complex [21] or determine them to be inadequate for evaluations on a scale smaller than the organizational or the facility level [22]. Others contend that the sustainability metrics that simultaneously address environmental, social, and economic aspects are ideal [23]. For example, nonrenewable energy use also encompasses social and economic concerns like pollutants and GHG emissions. One drawback of this metrics selection method is that it is inherently partial to environmentally based metrics. Specifically, any environmental issue can be rationalized as also being social, since only the environmental matters that create social problems tend to be recognized, but the reverse is not necessarily true (i.e., social issues do not always have an environmental component as well).

However, environmental metrics alone cannot comprehensively cover all aspects of sustainability. If this were the case, the distinction of separate categories for social and economic aspects in the three-dimensional definition of sustainability would be superfluous. Also, it is clear that the intention of including social aspects in the intergenerational and equality-based conceptions of sustainability [24, 25] was not merely to filter out environmental issues that have no social relevance. Many social concerns cannot be addressed through the proxy of an environmental metric. For instance, the injury and fatality rate of a production process does not have surrogate environmental metrics associated with it. And, unlike environmental issues, social goals (i.e., acceptable living standards for all of the world's populations) tend to be subjective, making it nearly impossible to reach consensus on these issues. There is more discussion on the policy implications of social sustainability in this chapter.

So, given these difficulties how can social metrics best be determined? The general methodology for developing these metrics is the same as will be outlined for metrics in general. However, to overcome the problematic nature of social metrics development, two main aspects should be emphasized. First, the overarching objectives (i.e., the general issues) which the social metrics aim to represent must clearly be articulated. For instance, Azapagic and Perdan [26] grouped social metrics into three broad categories based on the responsibilities that businesses have to communities and society. These are (1) human development and welfare which includes topics such as education and training, health and safety, and management competence; (2) equity which consists of wages and benefits, equal opportunity, and nondiscrimination; and (3) ethical considerations which cover human rights, cultural values, and intergenerational justice. Because social metrics are so contentious, it needs to be clear what their ultimate intention is.

A feasability study titled "Integration of social aspects into LCA" by the Task Force of the UNEP-SETAC Life Cycle Initiative on the integration of social aspects into LCA concluded that it is generally possible to transfer the LCA methodology to social aspects [27]. However they identified considerable hurdles in defining, modeling, categorizing, and characterizing social indicators. They also noted the difficulties that arise from the lack of availability of reliable data, and the subjectivity of social indicators (i.e., from different nations and cultures). Their past experience in sustainability assessments shows that social aspects that are related to products consist of three types: (1) effects at the level of resource extraction and upstream chains, (2) impacts upon consumers, and (3) indirect effects of product use upon society.

In addition to clearly identifying the objectives of social metrics, the scope of these metrics must also be explicitly established. Azapagic [28] distinguishes social metrics into two categories of micro-perspective metrics which relate to employees and macro-perspective metrics which relate to society at large. Indicator lists for social metrics have been developed by the Organization for Economic Co-operation and Development [29], the Commission for Sustainable Development [1], and the annual global environmental report "World Resources" [30] on a global/national level and by the Product Sustainability Assessment (Sect. 3.2.4) and Global Reporting Initiative [5] on a company/product level. The Lowell Center for Sustainable Production outlines a five-level framework for developing sustainability metrics, which increasingly grows in scope over time [31]. This framework can be particularly helpful in developing social metrics.

First, metrics related to regulation compliance and industry-standards conformance must be considered for facilities. In the context of social metrics these represent the risks of actions or fines against businesses for their treatment of workers and community members. Examples of these types of metrics pertain to OSHA standards and recordable injury and illness rates. At the next level are metrics related to facility performance and material use. These metrics benchmark the efforts put forth by producers. Some illustrations of these metrics include the number of community outreach programs; amount of employee training; and probability of workstation hazards. Next, metrics that measure the actual effects of facilities are identified. For social metrics these evaluate the impacts of production on workers and communities. Metrics that are representative of this type consist of concentration of specific contaminants in ground/surface water; percent of days with poor air quality as a result of a facility production; worker hearing loss; and employee retention rates. The subsequent stage of metrics relate to the supply chain and product life cycle. These metrics measure the impacts beyond the scope of production including those by suppliers and consumers. Typical metrics of this type relate to the location of sourced materials (e.g., locally, or from developing countries) and the working conditions at supplier facilities. Finally, metrics for overall sustainable systems are chosen. These metrics contain some sort of information about the larger context in which operations take place. Metrics that are characteristic of this type are income disparity within company and compared to local community and industry, and community quality of life.

However, these proposed social metrics methodologies still do not address the problem of scope posed by Reich-Weiser et al. [22]. Most social metrics measure impacts at the facility or the organization level. Developing techniques to proportion these measures to the process or the product level can be difficult. For products, impacts can be scaled by dividing the total facility impacts by the production rate to obtain a per product impact. However, this measure can be complicated when a facility creates more than one product. At the machine or the process level, there are already several metrics which relate directly to these impacts, like the amount of noise or injuries associated with a workstation. However, the question of how to allot the impacts outside the immediate scope of the process still remains. For instance, should quality of life metrics such as poverty and access to health care be attributed to a machine, and if there are multiple processes how are these determined? These issues clearly must be resolved to adequately address social metrics at these levels.

3.2.4 Multiple Metrics

Since most authors who propose metrics for sustainability are attempting to create a universal set, they tend not to be explicit in outlining a methodology to constructing such sets. Neely et al. [32] note that very little guidance is provided on how appropriate metrics can be identified. Much of the work which provides a framework for selecting sustainability metrics focuses on the various aspects of goal definition and metrics components. Both of these aspects are necessary to arrive at a comprehensive sustainability metrics set. Reich-Weiser et al. [22] use a methodology for metrics selection for manufacturing which follows the first steps of the ISO 14040 standards on LCA [33]. The steps of their methodology include (1) defining the goals of the assessment (sustainability concern); (2) choosing a metric type; (3) determining the manufacturing scope; and (4) determining the geometric scope.

Ragas et al. [34] outline a six-step process for the development of a sustainability indicator. First environmental problems are classified into themes and subthemes. An example of a theme is climate change, for which possible subthemes are greenhouse effect and depletion of the ozone layer. Next determinations of the "environmental space," or the thresholds for consumption and emissions given current stocks and capacity, are calculated for each subtheme. From these figures the environmental space for the relevant geographical scope is extracted. In addition the real environmental impacts are calculated using LCA methods. Then they are compared to the environmental space of the production scale to obtain a value known as the section indicator. This is done in a ratio where real impacts are used in the numerator so that any quotient larger than one gives an immediate indication of unsustainable practices. Next an overall indicator is calculated by multiplying each section indicator by a weighting factor, and then averaging the sum of all of them.

Veleva and Ellenbecker [31] propose an eight-step continuous-loop model for defining and measuring the sustainability performance of companies. The first three steps of their methodology relate to metrics selection, which consists of defining sustainable production goals and objectives, consistent with Lowell Center for Sustainable Production principles; identifying potential core and supplemental indicators; and selecting indicators for implementation. To achieve these goals, they also provide a five-level hierarchy for metrics development. The top tier of their hierarchy uses the three aspects of sustainability, which are further refined into six dimensions of sustainable production: energy and material use (resources); natural environment (sinks); social justice and community development; economic performance; workers; and products. For each of these aims, it is recommended that goals or targets particular to the organization be set. Goals are classified into five categories that correspond to evolutionary steps towards more sustainable production. The purpose of these categories is to represent the idea that organizations should begin with straightforward, easy-to-implement measures of compliance and resource efficiency and progress towards more advanced indicators, addressing environmental effects, social effects, and supply chain and life-cycle impacts. Metrics (which Veleva and Ellenbecker term indicators) can be generated to reflect goals and the current stage a company is at in this evolutionary process. However, they must be selected in accordance with the nine guiding principles for sustainable production presented by the Lowell Center for Sustainable Production. These principles are the following: (1) Products and packaging are designed to be safe and ecologically sound throughout their life cycles; services are designed to be safe and ecologically sound; (2) wastes and ecologically incompatible by-products are continuously reduced, eliminated, or recycled; (3) energy and materials are conserved, and the forms of energy and materials used are most appropriate for the desired ends; (4) chemical substances, physical agents, technologies, and work practices that present hazards to human health or the environment are continuously reduced or eliminated; (5) workplaces are designed to minimize or eliminate physical, chemical, biological, and ergonomic hazards; (6) management is committed to an open, participatory process of continuous evaluation and improvement, focused on the long-term economic performance of the firm; (7) work is organized to conserve and enhance the efficiency and creativity of employees; (8) the security and well-being of all employees is a priority, as is the continuous development of their talents and capacities; and (9) the communities around workplaces are respected and enhanced economically, socially, culturally, and physically; equity and fairness are promoted [35]. Lastly, the unit of measurement, type of measurement (absolute or unitized), period of measurement, and boundaries (i.e., the scope) of each metric must be decided upon [31].

Developed by the German Öko-Institut, the "Product Sustainability Assessment (PROSA)" is a method for strategic analysis and the evaluation of products and services. It is a methodology which is process driven and iterative and mainly based on well-established individual tools, e.g., LCA or LCC. Figure 3.1 shows the structure and sequence of work involved with the method [36]. The ecologic analysis is based on the ISO 14044 norm and the economic analysis is based on the work of SETAC. The social analysis is based on the work within the UNEP-SETAC Taskforce "Integration of Social Aspects into LCA" [27]. PROSA proposes a reduced list of 40 social indicators out of over 3,000 identified. In addition to the three dimensions the benefit of the product or the service can be measured based on proposed indicators for practical utility, symbolic utility, and societal benefit. Each of these four analyses can be aggregated to a single value which can be represented as a bar chart or a spider diagram.

The aim of the BASF tool SEEbalance is the quantitative evaluation of products and production processes regarding the ecological, economical, and social dimension

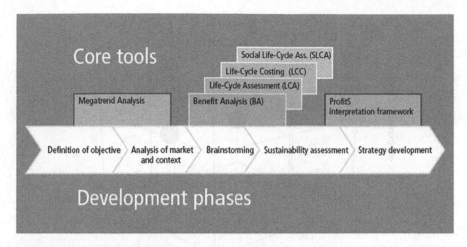

Fig. 3.1 Basic structure of PROSA [x] Copyright 2007 Rainer Grießhammer1 reprinted with permission

of sustainability [37]. The ecological assessment is based on ISO 14044 with an aggregation to the five categories of resource consumption; energy consumption; emissions to air, water, and soil; ecotoxicity potential; and land use [33]. The final ecological assessment is calculated based on weightings for these five categories. The economic assessment is based on an overall cost analysis including material and energy flows and all relevant secondary processes. The social impacts are based on the five stakeholder groups: employees, international community, future generation, consumer, and local and national community. The results of the analysis are represented in the three-dimensional diagram SEEcube; see Fig. 3.2. Users can evaluate various alternatives on the basis of the three axes: environmental burden, costs, and social influences. Moving to the right and upper corner of the cube indicates higher socio-eco efficiency. Lower to the left indicates lower socio-eco efficiency.

Ford of Europe's Product Sustainability Index (PSI) is a sustainability management tool which was developed to be directly used for engineering, i.e., not by sustainability or life cycle experts. PSI is based on eight indicators reflecting environmental (Life Cycle Global Warming Potential, Life Cycle Air Quality Potential, Sustainable Materials, Restricted Substances, Drive-by-Exterior-Noise), social (Mobility Capability, Safety), and economic (Life Cycle Cost of Ownership) vehicle attributes. PSI does not reduce the evaluation to a single value. PSI is integrated in Ford Product Development System and included in the companies' "Multi Panel Chart" where all product attributes are tracked during the development process [38].

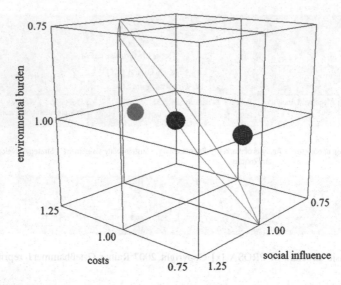

Fig. 3.2 Visualization with the three-axis diagram, after SEEcube

---

3.3   Overview of LCA Methodologies

3.3.1   Overview of Three Types of LCA

Environmental LCA is necessary to determine the environmental impacts associated with a process, product, or service. LCA is generally described as a systematic analysis of the material flows associated with every stage of a product's existence throughout material extraction, manufacturing, distribution, use, and end of life; however, in practice any number of these stages might be left out of the analysis. The ISO 14040 series of standards define LCA guidelines and establish four stages to an LCA: (1) goal and scope definition; (2) inventory assessment; (3) impact assessment; and (4) interpretation of results [33].

The first step is to determine the goal and analysis boundary. A boundary is defined as the set of activities within the total product's life cycle that will be considered. The boundary can be set to include all manufacturing operations, a factory, a machine tool, or a geographic region. For certain environmental metrics it is necessary to determine the total impact associated with a product or a service globally. This is the case for GHG emissions as they contribute to global climate change regardless of emission location and insight can be gained from a global perspective. However, a global assessment is not necessarily appropriate for a metric such as water, because the total water use associated with a product's manufacture across multiple locations provides no indication of regional environmental damage from overuse. For the case of water, a "gate-to-gate," or region-specific, analysis might be most appropriate.

The second step to an LCA, inventory assessment, is arguably the most time-consuming, where detailed data collection is required across a range of processes to obtain a complete picture of environmental impact [33]. Life-cycle inventories can be obtained using one of the three general methodologies: process LCA, input–output LCA, or a hybrid combination of process and input–output.

Process LCA is the most common method for inventory assessment. Process LCA consists of methodically analyzing material flows at every stage of the life cycle to understand precise consumption and emission values. In many cases, the work of previous researchers on certain materials or processes is included to complete the analysis. For situations where the desired analysis boundary is finite, as it will be for water consumption, a process approach must be utilized. However, when a comprehensive supply-chain analysis is desired, process LCA has an inherent boundary definition problem. This is because a supply-chain is inherently infinite, as everyone is using something from someone else to accomplish his or her piece of the chain, and every component of a system simply cannot be accounted for by the LCA practitioner given time and cost constraints.

The boundary problem is solved by using an EIOLCA approach. EIOLCA utilizes EIO tables and industry environmental data to construct a database of environmental impact per dollar of production in a given industry [39]. There is a large setup cost in creating the EIOLCA database; however it is relatively straight-forward to use once in place as financial data can be mapped to EIOLCA data directly. This method solves the boundary problem of process LCA because the EIO tables capture the interrelations of all economic sectors; however, Input–output LCA has the problem of providing only aggregate industry-level data.

While process LCA can provide a detailed analysis of specific process flows, IO LCA is able to quickly capture the interrelations between all sectors of the economy [40–42]. The scope of process LCA is limited by time and data, whereas the specificity of IO LCA is limited by the granularity of the IO table. Hybrid LCA can be thought of in one of the two ways: either process data is augmented with IO data to provide a comprehensive analysis or IO data is modified by process data to provide specificity.

Suh and Huppes [43] define three types of hybrid analysis: tiered hybrid, I–O hybrid, and integrated hybrid. A tiered hybrid methodology uses process data for upstream use-phase and end-of-life calculations, and supplements the downstream calculations with I–O data. In a tiered hybrid methodology it is unnecessary to combine the process and IO data in a single matrix, although it is possible. The IO hybrid methodology essentially disaggregates an existing category in the IO matrix into multiple more specific categories—the resulting table is then used as it would be in the tiered hybrid approach. Integrated hybrid LCA establishes a matrix representation of all system flows and utilizes the IO portion of the matrix to complete process LCA cutoffs.

3.3.1.1 Impact Assessment
Ultimately, the goal of environmental metrics is to quantify the impact of various activities on the environment.

Life cycle impact assessment, as a process for evaluating the environmental effect of emissions in the life cycle inventory, is developed as an analytical tool to support decision making in the environmental control and management practices. Current impact analysis methods are all developed by following the following three steps: (1) define impact categories and link the inventory data to the impact categories, (2) normalize the characterized impact category values to eliminate the embedded dimension differences among various impact categories to prepare for weighting which is the only way to combine the impact assessment results into one indicator to support decision making, and (3) use weighting and grouping methods to combine the impact analysis results into a single value for comprehensive benchmarking.

The life cycle impact result is calculated through the following formula:

$$\text{impact} = \sum_k \left(V_k \times \frac{\sum_s \text{CF}_s \times E_s}{R_k} \right), \tag{3.8}$$

where CF_s is the characterization factor for emission material s, E_s is the amount of emission material s, R_k is the normalization value of impact category k, and V_k is the weighting factor of impact category k.

There are dozens of impact assessment methods being developed worldwide such as Eco-Indicator 99, the impact assessment method developed by PRe consultants in the Netherland [46] and TRACI, the impact assessment method developed by the US EPA.

Some impact categories are agreed upon and commonly appear in LCA, such as acidification, global warming, eutrophication, ozone depletion, etc. But, there are also some categories that don't appear across all methods, such as land use and summer smog. It has been noted that the flexibility of impact categories undermines the validity and consistency of environmental impact assessment. As there is no standard impact assessment method currently acceptable for use in the CLA community, lack of impact assessment methods has become one of the major barriers prohibiting wide use of LCA.

3.3.1.2 Risk Assessment

Impact assessments are informed by risk models developed for particular emissions to air, land, and water. Substances used in product design and manufacturing have significant impact on the environment and human health. The potential impact of chemicals on human health is typically assessed by using a risk assessment method [51]. The risk assessment paradigm was established by the National Research Council [52] report, initially for regulatory decision making, and later extended by many researches to a wide range of applications including environmental and human health impact assessment as described in [53–57].

Currently risk assessment is carried out through a standard four-step process: hazard identification, exposure assessment, dose–response assessment, and risk

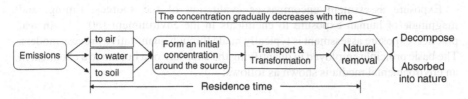

Fig. 3.3 Fate and transport of a pollutant in the environment

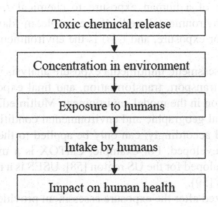

Fig. 3.4 Human health impact process of toxic chemicals

characterization [51]. Toxic chemicals, as released into an environmental media (air, water, land) form a concentration initially around the source and then are transported to a wide range of areas through natural transport processes such as advection, dispersion, diffusion, etc. During the process of transport, these chemicals may undergo a series of transformation processes, for example, by reacting with existing substances in the environmental media, or being subject to the photolysis of sunlight, or phase changes between different environment media. Finally these released chemicals would be either absorbed into nature or decomposed after a period of time through such transformation processes. The time period starting from the initial release until the final removal out of the environment is defined as the persistence residence time of a pollutant in the environment, which varies from a few seconds to hundreds of thousands of years, depending on the material properties of the released chemical and the environmental conditions the chemical is released into. The environmental transport and transformation process of a chemical release are shown in Fig. 3.3.

During its persistence (or residence) time, a released chemical could cause potential impact on human health through a number of exposure pathways (air, water, soil, food, etc.) and various exposure routes (inhalation, ingestion, dermal uptake, etc.) on an exposed population. The human health impact process of a toxic chemical through environmental media is shown as follows in Fig. 3.4.

Exposure assessment encompasses evaluation of the sources, timing, and magnitude of human exposure to chemicals in the environment [50, 51]. In real practice, exposure assessment is typically made through quantitative process models. The basic principle employed in quantifying the exposure of a human to a chemical in an environmental media is shown as follows:

$$D_i = K_i \times \int_0^T C_i(t)dt, \qquad (3.9)$$

where D_i is the dose of a human exposure to chemical i, K_i is the human's consumption rate of environmental media (for example, m^3/day for air inhalation), T is the time period for exposure, and $C_i(t)$ is the environmental concentration of chemical i at time t.

In a quantitative assessment, multimedia exposure analysis models are typically used for tracking the transport, transformation, and final exposure of a chemical release to the population in the model environment. Multimedia exposure analysis models rely on regional geographic and environmental conditions for the fate and transport analysis, and accordingly, can only be applied to the specific region for which the model is developed. For example, CalTOX is a multimedia exposure assessment model developed for the US region [58]; USES is a model developed for European applications [59].

The risk characterized after the exposure assessment provides an estimate of the adverse health effect on the exposed population from a toxic chemical release. Risk of a human exposure is typically assessed through the ratio between the final dose (as calculated by 3.14) and a reference dose value which is a quantitative indicator obtained through dose–response assessment:

$$R_i = \frac{D_i}{V_i}, \qquad (3.10)$$

where R_i is the risk of chemical i, D_i is the final dose of chemical i, and V_i is the reference dose value for chemical i.

3.3.2 Material Flow Analysis

Material Flow Analysis (MFA) is a methodology developed for quantitative analysis of material flows into and out of a subject system. MFA is a material accounting procedure widely employed in the study of industrial ecology topics [44]. MFA generally employs a material balance approach for the analysis of material flows within a target system. The MFA target could be a selected substance (a chemically defined element or a compound such as carbon dioxide), a material (natural or technically transformed matter that is used for commercial or noncommercial purposes such as platinum), a product (such as a computer), an industrial sector

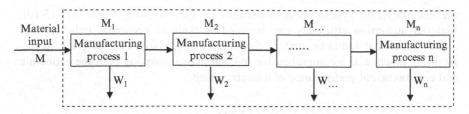

Fig. 3.5 A single material flow analysis in a manufacturing system

(such as semiconductor manufacturing industry), or an economy (such as the US economic system) [45]. MFA can be conducted in various scales at international, national, or regional target systems.

The quantitative analysis models developed based on MFA can aid decision-makers to gain deep insights into the target system to systematically understand the material flows, sinks, and interactions within the system. MFA is particularly effective for identifying hidden material flows and waste generation mechanisms in the target system, and to meet the needs of sustainable development for minimizing material consumption and waste generations from industrial production activities.

MFA, when applied to a manufacturing system, can be used as a system analysis tool to provide comprehensive decision support for green manufacturing through such efforts as quantifying the input materials and output waste, quantifying material utilization efficiency, identifying improvement opportunities, etc. The detailed application of MFA in a manufacturing system is described in the following.

A manufacturing system typically comprises a series of manufacturing processes. Each process has its material requirements and consumes a certain amount of materials for making products during the production process, with some amount of waste generated from the material used in the manufacturing process.

A simple demonstration of the MFA principles on a single material flow within a manufacturing system is shown in Fig. 3.5. The manufacturing system comprises n manufacturing processes with a total amount of material input M. For each process, an amount of M_i is consumed to make products (flows into the products) and an amount of W_i waste is generated from the manufacturing process i.

According to the material balance principle, the total amount of input into the manufacturing system equals to the total amount being used for product-making and the total amount of waste generated from various manufacturing process, as shown in the following expression:

$$M = \sum_{i=1}^{n} M_i + \sum_{i=1}^{n} W_i. \tag{3.11}$$

Through MFA, the material utilization efficiency of any given material within a manufacturing system can be identified. Material utilization efficiency is defined as the ratio of the amount used for product-making and the total amount of input into the manufacturing system, as demonstrated by the following expression (3.10). Material utilization efficiency can be used as both an economic indicator and an environmental indicator, and, accordingly, is popularly used in green manufacturing practice as a comprehensive indicator for improving both the economic and environmental performance of manufacturing:

$$\eta = \frac{\sum_{i=1}^{n} M_i}{M}. \qquad (3.12)$$

In the above analysis, only a single material input and its flow within a manufacturing system are considered. However, a manufacturing system usually requires multiple material inputs as both working materials for making products and supplemental materials to aid in manufacturing operations. For a comprehensive analysis of material flows within a manufacturing system, the following matrix is derived for expressing the consumption of multiple materials within a manufacturing system. Here we define a total of m kinds of materials as inputs to a manufacturing system which has n manufacturing processes:

$$\begin{bmatrix} M_1 \\ M_2 \\ \vdots \\ M_m \end{bmatrix} = \begin{bmatrix} M_{11} & M_{12} & \cdots & M_{1n} \\ M_{21} & M_{22} & \cdots & M_{2n} \\ \vdots & \vdots & \vdots & \vdots \\ M_{m1} & M_{m2} & \cdots & M_{mn} \end{bmatrix} \begin{bmatrix} 1 \\ 1 \\ \vdots \\ 1 \end{bmatrix} + \begin{bmatrix} W_{11} & W_{12} & \cdots & W_{1n} \\ W_{21} & W_{22} & \cdots & W_{2n} \\ \vdots & \vdots & \vdots & \vdots \\ W_{m1} & W_{m2} & \cdots & W_{mn} \end{bmatrix} \begin{bmatrix} 1 \\ 1 \\ \vdots \\ 1 \end{bmatrix}. \qquad (3.13)$$

In summary, MFA is a system tool which offers a comprehensive view of complex material flows within a manufacturing system and can be used to support decision making about the green manufacturing efforts on improving material utilization efficiency and waste minimization.

3.3.3 Energy Flow Analysis

Energy flow analysis (EFA) is a methodology developed for tracking and understanding the energy flows within a complex system, often for decision support in minimizing the energy consumption of the target system to reduce its environmental impact from fossil fuel energy use. EFA is very similar to the MFA as described in previous section in terms of modeling structure and data acquisition.

In a manufacturing system, energy is universally needed to drive machines and operate manufacturing processes. The energy flows within a manufacturing system could be modeled by using an equipment-centric approach like EnV-S as described in [17]. In this way, the total amount of energy consumed in a manufacturing system equals the sum of energy consumed by each manufacturing facility, as shown by the following expression:

Table 3.1 Average air emission factors for electricity generation in the United States

Emission	Amount (g/kwh)
CO_2	642
CH_4	2.42
NO_x	1.56
N_2O	0.0048
Pb	0.000115
PM	0.879
SO_x	2.46
CO	0.115
HCl	0.148
HF	0.0157
Hg	0.0000213

$$Q = \sum_{i=1}^{N} P_i \times t_i, \qquad (3.14)$$

where Q is the total amount of energy consumed in a manufacturing system, P_i is the power demand of device i, t_i is the operating time of device i, and N is the total number of devices in the manufacturing system.

Energy flows are of much concern in a manufacturing system mainly because of the costs and environmental emissions associated with current energy supply systems [49].

For assessing the environmental impact of energy use, the total amount of energy consumed is typically used as an indicator by the manufacturing industry. Quite frequently, the energy-induced emissions are also used for establishing an emissions inventory of manufacturing, especially in LCA. The emissions from energy use could be calculated through emission factors based on a unit process model [49]. The calculated emission factors per unit energy consumption, based on a kilo-watt-hour, vary slightly from state to state in the United States due to different electricity supply patterns and energy sources [49]. Table 3.1 below shows the average of air emissions per kWh of electricity generated from the 50 states in the United States [49].

For a total amount of energy consumption, Q, the manufacturing emissions inventory could be established by using the following matrix for multi-emission calculations. Here we assume that there are a total of k types of emissions generated from the energy consumed for electricity productions. E_i is the emission material; F_{ii} is the emission factor:

$$\begin{bmatrix} E_1 \\ E_2 \\ \vdots \\ E_k \end{bmatrix} = Q \begin{bmatrix} F_{11} & 0 & \cdots & 0 \\ 0 & F_{22} & \cdots & 0 \\ \vdots & \vdots & \ddots & \vdots \\ 0 & 0 & \cdots & F_{kk} \end{bmatrix} \begin{bmatrix} 1 \\ 1 \\ \vdots \\ 1 \end{bmatrix}. \qquad (3.15)$$

3.4   Metrics Development Methodologies

3.4.1   Ecological/Cost Metric Choice Model

A 4-part methodology has been proposed to determine appropriate environmental metrics as part of a sustainable manufacturing strategy [74]. Metrics are identified based on the particular concerns being addressed in the sustainability study. Colloquially, we are looking for "the right tool for the right job"' as it is difficult to conceive an absolute "best"' metric for sustainable manufacturing. Additionally, this methodology is intended to be flexible and modular over time, which is important given that the effectiveness of the metric is determined only by its usefulness in a specific context. Determination of appropriate metrics is inherently influenced by current "social value, knowledge horizons, and individual perspectives" [60].

It should be noted that this methodology follows the ISO 14040 standards on LCA [33]. The four main steps of LCA are goal and scope definition, inventory analysis, impact assessment, and interpretation of results. With this methodology we are essentially performing the first step of ISO14040 as it is relevant to metric selection. Steps 1 and 2 define the metric's goal, while steps 3 and 4 determine scope.

Step 1: Goal Definition—Determine the goal of the assessment. This first step requires an understanding of the sustainability concerns driving the effort. This means that the metric selection needs to be driven by the objective of the sustainable manufacturing strategy. Additionally, at this stage the functional unit for the assessment should be determined.

Furthermore, if a technology is new, or requires the processing of new materials that are poorly understood, then a comprehensive sustainability assessment employing a suite of metrics may be necessary. However, if we are studying specific impacts or the consumption of particular resources, then it is adequate to only highlight these concerns. Care should be taken not to overly simplify the assessment goals; however with enough information, simplification and scope reduction at this stage can be useful in reducing the time and costs needed for the sustainability assessment.

Step 2: Goal Definition—Choose a metric type. Generally, metrics for manufacturing decision making can be classified as either "cost" or "sustainability" indicators. "Cost" metrics can also be characterized as "green" rather than "sustainability" metrics. Here, these categories are further broken down into four distinct metric types.

The first two metric types are analogous to familiar cost metrics. First are the *intensity* metrics, which indicate the cost per functional unit. Second are *return on investment* metrics that indicate the percent savings of a particular investment relative to the input required for the investment.

The third and fourth metric types are based on sustainability concerns relative to resource availability. Use of resources that are considered "renewable" can be characterized by an *availability factor*, which indicates consumption relative to replenishment rates. The availability is the "amount of resource use" relative to the "total resource availability." This is comparable to machine tool availability metrics used in measuring the efficiency of manufacturing systems.

Decision metrics for nonrenewable resources is an area requiring further research; however one way to quickly understand the risk associated with using nonrenewable resources is by calculating the *time remaining* of the resource given current consumption patterns and available reserves. Because this value does not enable decision making at all levels of production, it highlights the need for metrics to understand nonrenewable resource consumption.

Step 3: Scope Definition—Determine the manufacturing scope of the assessment. While it is always important in the development of green technologies to consider the life cycle of the technology—which includes material extraction and conversion, industrial facilities usage, process consumables usage, manufacturing process impacts, supply chain and transportation impacts, product use, and end of life— decision making often must occur on a smaller scope within the larger. Decision making in a manufacturing enterprise can take place at many different levels; therefore the scope of application should be understood when using the metric formulations given above. For example, when investigating the manufacture of a product, the eventual use and end of life of that product need not be considered, unless decisions during the manufacturing stage have an impact on the use phase, or the end of life.

The following levels of analysis scope are identified:

Machine Tool Scale: At this scale, decisions specific to one machine tool or a small family of tools are taken. The decisions are usually made regarding the fundamental process technology. Control of lubrication systems and MQL is an example of decisions for sustainable manufacturing at this level. Metrics at this scale reflect the functionality of the machine tool (e.g.: emissions per minute or energy consumption per part). The "ripple" effects of decisions taken at this scale must be considered by analyzing subsequent manufacturing operations (for example: using MQL could necessitate additional cleaning operations).

Line Scale: This scale includes the set of machine tools and support equipment that are logically organized into a manufacturing line or cell. Final and intermediate products are created at this scale and metrics need to be relevant to the entire scope of this scale including support equipment and machinery.

Factory Scale: Here the entire factory is incorporated and metrics at this scale need to capture the impact of the facility itself on the environment. For example, the total water and energy consumptions of a semiconductor facility must take into account HVAC and clean-room systems [61].

Supply Chain Scale: This level looks at the manufacturing enterprise including its entire supply chain. At this scale, metrics need to be selected to capture the interrelationships between discrete geographical entities in the system. The effect of the complex transportation and communication networks prevalent in manufacturing systems also needs to be accounted for [62]. The metrics at the factory and supply chain scales also need to comply with local, national, and international standards because an economic cost can be associated with these. For example, in the United States emissions not known to be an environmental hazard at the time can be later subject to large fines through the Superfund program [63].

Life Cycle Scale: The final level goes beyond the supply chain to include product use and end-of-life decisions. This will incorporate supply chains associated with consumables throughout the use phase, operational and maintenance impacts, as well as end-of-life reverse logistics, recycling, reuse, and disposal.

Note that each scale incorporates the effects of lower scales [64]. For example, the supply chain scale includes all the factories throughout the system, plus transportation and logistics. The factory scale includes all of the product lines as well as extraneous factory requirements such as HVAC and overhead. The line scale includes all machines in the line plus transport between machinery. Given the complexity of decision making across these scales, it is critical to clearly identify at which scale (or scales) the sustainability metrics are going to be applied. It may not be possible to select a metric that is relevant or applicable across all the scales. For example a metric of local water availability cannot be readily applied across a global supply chain.

Metrics at the lowest scale tend to be customized for specific process technology (such as consumable consumption rates) and local environmental conditions for sustainability. Metrics at the higher scales can be broad-based enough to be applied at the lower scales (such as carbon emissions or energy consumption), but not necessarily vice versa.

Step 4: Scope Definition—Determine the geographic scope of the assessment. While in some cases the manufacturing scope defines the geographic scope of the assessment, this is not always necessarily true. For example, a sustainability metric based on energy use can be related to either global energy resources or local energy infrastructure capacity. Depending on the goal of the assessment, the appropriate geographic scope can be determined.

Choosing a metric requires understanding the geographic range of the environmental concern. Environmental impacts may be highly localized or globalized. For example, greenhouse gas emissions can affect global climate change regardless of where they are released. However, if electricity supply is scarce in one location, excessive use of electricity elsewhere is neither harmful nor helpful to the local scarcity.

3.4.2   Decision Tree Model for Equipment Investments

In this section, a concept for the integration of ecological and social criteria in the economic investment evaluation process is described. In order to include as well the changeability of the equipment the basis for the evaluation is a decision tree model (Fig. 3.6). The method consists of five phases, which can be applied in an iterative manner in order to detail the result [65].

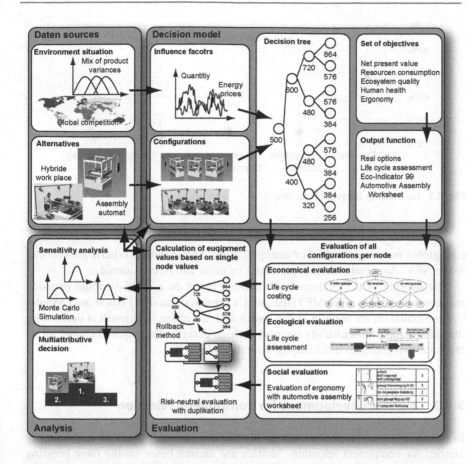

Fig. 3.6 Methodology for the multi-attributive evaluation of assembly equipment, from [65]

The first phase of the method is dedicated to the development of the decision tree for the designated product. Applying the scenario method by Gausemeier et al. [66] key influence factors will be identified and future developments will be estimated. In adaptation of [67] a primary influence factor, e.g. demand, will be selected and a decision tree will be modeled based in the future developments of this factor. The figure shows a binominal not recombining tree which allows an up and down movement of each note from one period to another.

Depending on the identified key influence factors, relevant technical possibilities for the equipment to adapt to the changing environment will be determined in the second phase.

During the third phase the calculation of the decision tree based on the real options approach is conducted. Within this calculation process the economic, ecological, and social criteria are treated separately. The ecological criteria are classified and characterized according to internationally accepted impact categories,

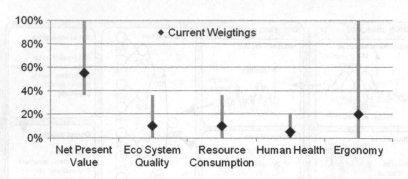

Fig. 3.7 Insensitivity Intervals, from [65]

e.g., global warming potential or eutrophication. The economic criteria are reduced to the NPV. As social criteria ergonomics or noise can be considered.

The ranking of the evaluated alternatives based on a multi-attribute decision making method, e.g., PROMETHEE, and a sensitivity analysis based on a Monte-Carlo Simulation are the aim of the fourth and last phase. The result is verified by the calculation of insensitivity intervals, which describe the required change of an attributes weighting to influence the overall ranking (Fig. 3.7).

3.4.3 Metrics development for component selection

In this section, we propose a six-step algorithm to develop a set of sustainability metrics for component selection. Metrics are chosen based on the most pressing sustainability issues specific to the application being addressed. While the concerns of sustainability do not generally change over time, some matters become more urgent, while others are alleviated through proactive measures. In addition, since metrics serve as proxies for issues, other measurements, which are more representative, may be discovered over time. For instance, ozone depletion continues to remain a key issue of global warming. However, with global efforts to phase out CFCs, as a result of an increased awareness of their role in ozone depletion, other issues of global warming have come to the forefront.

The methodology proposed below relies on the insights of decision makers (with the input of stakeholders) to select the most appropriate metrics for sustainability. One danger in the selection process is the inherent subjectivity of individuals. However, given the nature of sustainability assessments, we do not believe that this should be cause for concern. We begin by assuming that the intention of creating a sustainability metrics set is to obtain an accurate representation of sustainability. In addition, we further assume that there are two general motivations for conducting a sustainability evaluation: (1) to help an organization identify opportunities to improve their operations (i.e., for internal benchmarking) or

(2) to communicate the sustainability performance of an organization to outside parties such as consumers, regulatory entities, or other businesses. Given this assumption, if a practitioner were to attempt to manipulate the metrics selection process with the aim of obtaining an artificially high sustainability score, this would be inherently self-defeating. In the long run, such inadequate sustainability assessments will result in criticisms from the public as well as a loss of competitive advantage.

There are many benefits that an organization can gain from an accurate assessment of their sustainability. Azapagic [68] provides an extensive list of the opportunities that are afforded by sustainable practices, as well as the threats that result from a continued lack of concern for sustainability. These advantages relate to improvements in knowledge, efficiency, costs, and reputation. Assessments give an organization data on their own environmental performance, resulting in the knowledge of where improvements can be made and how best to refute criticisms. One possible area for improvements is in efficiencies, specifically in material and energy use, processes, or technologies. Another area for possible advances is in compliance with existing and forthcoming legislation. In becoming more sustainable, an organization may even be able to progress to a leadership position, allowing them the potential to help influence the legislative process. All of these potentialities can result in reductions of operational costs. Additional savings can be gained through reduced fines, lower insurance premiums, and even decreased "green" taxes (such as carbon taxes). A less tangible benefit that can be gained is an improvement in reputation. This trust that a company receives from consumers and communities can result in increased sales, the ability to penetrate new markets, and the possibility to attract talent to their organization.

As previously mentioned, the algorithm proposed here follows the implicit process taken by numerous authors in constructing sustainable metrics sets. Therefore the technique is not necessarily novel, but rather provides a clear definition of the procedure already used to aggregate relevant metrics; select those that are most pertinent to the application; and ensure that they are properly measured. This approach is general and can be used to generate sustainability metrics for any application. An overview of the metrics selection process is outlined below.

Step 1: Assume that the overall goal of the assessment is sustainability, which comprises a three-dimensional (social, environmental, and economic) set of considerations. This step constructs the uppermost level of the sustainable metrics hierarchy. It will also help guide the selection of subsequent steps ensuring that all aspects of sustainability have been comprehensively covered. It should be noted that at any point in time the individual elements of the hierarchy may not conceptually fall into any one distinct category of the level that preceded it. For instance, categories from the next step of the hierarchy may not exclusively fall into only one of the categories of social, environmental, or economic considerations. Despite this, all elements should be assigned to their most relevant grouping to ensure that each level is assembled as completely as possible, and guarantee that no aspects are overlooked. Thus, each dimension and subsequent clustering will be homogeneous and well separated.

Step 2: Determine the general issues that are relevant to sustainability. General issues are overarching concerns, which can comprise one or many components. Examples of general issues are global warming, resource scarcity, and pollution. The aim here is to get at the reasons why impacts are problematic and what prevents activities from being sustained indefinitely (i.e., the information that metrics data is supposed to represent). Authors often identify issues that in and of themselves are not cause for concern. While these categories may be the best proxy for the actual issues they represent, without clear recognition of their larger objectives, it is difficult to interpret the impact of improvements. To maintain objectivity and comprehensiveness in identifying general issues, it has been suggested that issue identification be carried out with a group of stakeholders who have diverse interests [69].

Step 3: Determine the specific sub-issues that are relevant to sustainability. Sub-issues differ from general issues in that they are issues whose impacts must in some way be measurable. Sometimes, general issues on their own have measurable impacts, and cannot be separated into sub-issues. In this case, the general issue and the sub-issue will be identical. The objective of aggregating sub-issues is to insure that the major contributors to any issue are identified. Examples of sub-issues are greenhouse gas emissions, raw material use, and emissions to water.

Step 4: Identify the specific issues that metrics will be related to. At this stage we are identifying what will possibly be measured, without the specific details of how these evaluations will be carried out. Examples of specific issues are recycled material content, renewable energy use, and amount of ground-level ozone created. This step will comprehensively aggregate all of the possible issues that metrics will directly be generated from.

Step 5: Determine the criteria to be used in metrics selection and apply them to select the specific issues that are most representative for the assessment. Criteria serve as bounds which metrics must fall into to qualify for use in assessments. They help to ensure that metrics are reasonable and meet the physical constraints of the evaluation. Criteria that are commonly used in developing sustainability metrics sets are the following [31, 70–73]:
– The number of metrics should be small and manageable.
– Metrics should be simple, and easy to understand and communicate.
– Measurements should be cost effective.
– Metrics that are relevant to sustainability should be chosen.
– Metrics should be general and scalable to diverse applications.
– The set should be useful for decision making.
– Metrics should be chosen to best ensure accuracy, reliability, and consistency.

This set is sufficient to generate a metrics set, although additional criteria might be useful if necessary. Once criteria have been identified, specific issues are chosen in accordance with these criteria and in a manner that comprehensively represent the general issues of the assessment. This set of specific issues will correspond to the core metrics that will be used for the assessment. The next step outlines how exact metrics are constructed.

Step 6: Detail the precise metric to be used, using the five metric dimensions of unit, type, period, system boundaries, and geographic scope. Metrics are created by turning the specific issues that have been selected into measurements. One objective in the creation of metrics is that they should be comparable amongst different applications. To achieve this, metrics must contain information across multiple dimensions. These dimensions are drawn from those that were provided by Veleva and Ellenbecker [31]: unit of measurement, type of measurement, period of measurement, and boundaries. The boundaries dimension is further subdivided into two categories, in accordance with the methodology suggested by Reich-Weiser et al. [74] and discussed in section 3.4.1, into the groupings of manufacturing and geographical scope. Examples of measurement units are numbers, weights, currency denominations, times, or percentages. Types of measurements are either absolute or unitized measures (i.e., a value per unit, product, or service). The period of measurement is the timescale being considered. The manufacturing scope is the portion of a production being considered (i.e., the machine tool, facility, line, supply chain, or life cycle). The geographical scope can be global, regional, or local. All metrics should entail each of these dimensions.

Several abstractions of this process can be made to further assist with its conceptualization. First, the steps of this process are shown below in Fig. 3.8, along with its associated dendrogram. Also, this process can be described through set notation, beginning with the following relevant nomenclature in Table 3.2.

Let $M = \{M_1, M_2, \ldots, M_n\}$ be a set of n entities, where M_i is equivalent to the specific issue represented by metrics i, and $i = 1, 2, \ldots, n$. Note that the cardinality of M is less than or equal to the cardinality of C, that is, $n \leq f$. Also, if we let E be a subset of M, then there exists a partition, $P_s = \{E_1, E_2, \ldots, E_s\}$ of M into s clusters, with a one-to-one correspondence between the elements of P and the elements of B. Also, the following holds.

3.5 Outlook and Research Needs

One of the vital requirements in assessing the sustainability of manufacturing processes or systems is defining suitable metrics. As we have seen in this chapter, there is a wide variety of metrics and methods to select from in sustainability assessment. This is a reflection of both the diversity of available frameworks in defining "sustainability" and the multiple interpretations and use-cases that are possible within a specific framework. Metric selection is considerably unambiguous when the scope of analysis is narrowly confined to a specific process or system. But as the complexity of engineered systems increases, metric selection becomes increasingly ambiguous as multiple metrics can potentially be applied in the analysis. Given that the competence of the analysis rests largely on the suitability of the metrics, the challenge in metric selection raises several concerns, including the following:

- How can the metrics capture the correct balance between immediate concerns of sustainability, versus concerns that will happen at a future point of time?

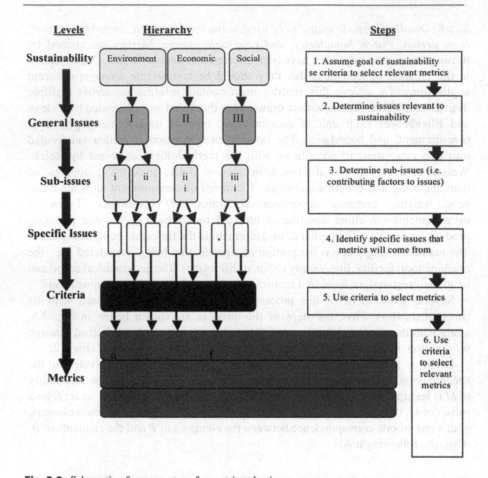

Fig. 3.8 Schematic of process steps for metric selection

Table 3.2 Notation for metrics hierarchy

Symbol	Meaning
G	Set of all general issues
B	Set of all sub-issues
C	Set of all specific issues
M	Set of all metrics
r	Number of general issues, $|G|$
s	Number of sub-issues, $|B|$
f	Number of specific issues, $|C|$
n	Number of metrics, $|M|$
t	Number of characteristics for each metric

- Can a single metric support multiple functional units of analysis?
- How do we compare two different metrics being applied to the same phenomenon? Can we define equivalencies between metrics?
- How do metrics that apply across over the entire life cycle of a process or a system relate to those that apply across a specific window of their life cycle?
- Can metrics that apply over specific windows of the life cycle of a process or a system be coalesced into metrics that apply over their entire life cycle?

In this chapter we saw a first attempt to address some of these questions. This chapter began by looking at metrics focused on the cost economics of decision making. These metrics are extensively applied in decision making, as they fully capture the complexity of purely economic considerations. Clearly, several of the environmental and ecologically focused metrics have taken inspiration from the economic metrics, because of familiarity, and because of the need to tie sustainability considerations strongly to economic considerations (which, looking into the future, will remain the chief consideration for most enterprises and governments).

Social metrics were then surveyed, and the challenge in quantitatively evaluating social concerns was illustrated by highlighting the multiple considerations that social metrics attempt to capture. We can clearly see here the limitations in selecting metrics for sustainability assessment from a social vantage; given the broad definition of "society," and the multiple, sometimes conflicting considerations that are deemed "valuable" to society, metrics can be useful only if the social framework is precisely defined, and those entities of value to the society are accurately characterized. Given the intangible nature of some social considerations, careful metric selection is even more important, as intangible qualitative metrics cannot be directly incorporated into conventional decision making frameworks (which predominantly tend to be very quantitatively focused).

The chapter then surveyed metrics that tie in multiple considerations, pulling together in ecological, social, and economic metrics. These meta-metrics can be more accurately understood as frameworks for decision making that balance potentially conflicting drivers. The metrics that were surveyed were all developed for specific application domains/industries, and are most valuable when they incorporate all the considerations and drivers that are important to the specific domain.

The chapter then discussed LCA methods to obtain the underlying inventory and impacts results required by green manufacturing and sustainability metrics. Process, EIO, and Hybrid LCA were discussed, and it was clear that choosing between these methods depends on how much information is available, and how precisely the process/system have to be modeled.

Finally, the chapter presented several approaches for metric development, which systematically build up the metric based on considerations of goal, scope, system boundaries, planning horizon, and system drivers. A common thread between these approaches is that they all define the metrics *only* based on their point of application—which brings us back to our initial point that metric selection goes a long way in determining the sustainability of a system. We can clearly see that it is very difficult to divorce the act of selecting metrics from the act of applying them.

Keeping these as independent as possible is important in ensuring that the metrics are unbiased and do not favor a particular solution or approach to system design.

Looking ahead, a key driver in developing sustainable manufacturing metrics has to be bias, or rather, the lack thereof. The proliferation of sustainability metrics makes it easy to "game" the analysis in favor of one solution (especially when the fuzzier metrics are applied). While this immediately defeats the purpose of the analysis, it does greater damage in the larger picture by weakening the intellectual rigor of sustainability engineering as a practice. In order to have robust tools to design and develop sustainable manufacturing systems, we require independent, unbiased, and flexible metrics. While we have made a lot of progress towards this goal, a lot more work is needed, and future research goals include the following:

- Develop tools that can compare metrics from different frameworks, and that can be used to build equivalencies between the frameworks.
- Develop a robust high-level framework for metrics that can be flexibly applied across multiple domains, and that take into account the different types of sustainability drivers.
- Develop metrics-of-metrics that can be applied in detecting the hidden (and not-so-hidden) biases present in existing sustainability metrics.

References

1. UN Commission on Sustainable Development (UNCSD) (2006) Indicators of sustainable development: guidelines and methodologies third edition. http://www.un.org/esa/sustdev/natlinfo/indicators/methodology_sheets.pdf. Accessed 27 Sept 2010
2. Parris TM, Kates RW (2003) Characterizing and measuring sustainable development. Annu Rev Environ Resour 28:559–586
3. Schweimer G, Levin M (2000) Life cycle inventory for the golf a4. Volkswagen AG, Wolfsburg. http://www.volkswagen-environment.de/. Accessed 17 Sep 2010
4. Brealey R, Myers SC (2003) Principles of corporate finance. 7th edn (Int Ed) McGraw-Hill, New York
5. Global Reporting Initiative (GRI) (2006) Sustainability reporting guidelines, Version 3.0. http://www.globalreporting.org/NR/rdonlyres/ED9E9B36-AB54-4DE1-BFF2-5F735235CA44/0/G3_GuidelinesENU.pdf. Accessed 17 Sep 2010
6. Wemhoner N (2005) Flexibility optimization to increase the capacity utilization in automotive body shell assembly. PhD thesis, WZL Aachen
7. VDI 2884 (2005) Purchase, operating and maintenance of production equipment using Life Cycle Costing (LCC). Beuth, Berlin
8. Woodward D (1997) Life cycle costing-theory, information acquisition and application. Int J Proj Manage 15(6):335–344
9. Kaufman RJ (1970) Life cycle costing: a decision making tool for capital eqmpment acquisition. Cost and Management, March/April: 21–28
10. Rebitzer G, Hunkeler D (2003) Life cycle costing in LCM ambitions, opportunities, and limitations: discussing a framework. Int J Life Cycle Assess 8(5):253–256
11. Bengtsson J (2001) Manufacturing flexibility and real options: A review. Int J Prod Econ 74:213–224
12. Black F, Scholes M (1973) The pricing of options and corporate liabilities. J Polit Econ 81:637–654
13. Copeland T, Antikarov V (2001) Real options: a practioner's guide. Texere, New York

14. Luehrman TA (1997) What's it worth: a general manager's guide to valuation. Harv Bus Rev 5–6(3):132–142
15. Luehrman TA (1998) Investment opportunuties as real options. Harv Bus Rev 07–08:51–67
16. Luehrman TA (1998) Strategy as a portfolio of real options. Harv Bus Rev 76(5):89
17. Krishnan N (2003) Design for environment (DFE) in semiconductor manufacturing. Ph.D. dissertation, University of California, Berkeley
18. Yuan CY, Zhang T, Rangarajan A, Dornfeld D, Ziemba B, Whitbeck R (2006) A decision-based analysis of compressed air usage patterns in automotive manufacturing. J Manuf Syst 25(4):293–300
19. Lundie S (1999) Life cycle assessment and decision theory—a praxis oriented product assessement based on societal values. Springer, Berlin
20. Geibler J, Liedke C, Wallbaum H, Schaller S (2006) Accounting for the social dimension of sustainability: experiencs from the biotechnology industry. Bus Strategy Environ 15(5):334–346
21. Araujo JB, Oliveira JFG (2008) Proposal of a methodology applied to the analysis and selection of performance indicators for sustainability evaluation systems. In: Curran R, Chou SY, Trappe A (org) Collaborative product and service life cycle management for a sustainable world, 1st edn. Springer, Germany
22. Reich-Weiser C, Dornfeld DA, Horne S (2008a) Environmental assessment and metrics for solar: case study of SolFocus solar concentrator systems. Proceedings of the IEEE PV specialists conference, San Diego, CA
23. Sikdar SK (2003) Sustainable development and sustainable metrics. AIChE J 49:1928
24. Stymne S, Jackson T (2000) Intra-generational equity and sustainable welfare: a time series analysis for the UK and Sweden. Ecological Econ 3:219–236
25. Okereke C (2006) Global environmental sustainability: intragenarational equity and conceptions of justice in multilateral environmental regimes. Geoforum 37:725–738
26. Azapagic A, Perdan S (2000) Indicators of sustainable development for industry: a general framework. Trans IChemE Part B Proc Safe Environ Protect 78(4):243–261
27. Griesshammer R, Benoît C, Dreyer LC, Flysjö A, Manhart A, Mazijn B, Méthot AL, Weidema B (2006) Feasibility study: integration of social aspects into LCA, Freiburg, Germany. http://www.estis.net/includes/file.asp?site=lcinit&file=2FF2C3C7-536F-45F2-90B4-7D9B0FA04CC8. Accessed 17 Sep 2010
28. Azapagic A (2004) Developing a framework for sustainable development indicators for the mining and minerals industry. J Cleaner Prod 12(6):639–62
29. Organisation for Economic Co-Operation and Development (OECD) (2009) OECD Factbook 2009: economic, environmental and social statistics. http://masetto.sourceoecd.org/vl=902364/cl=22/nw=1/rpsv/factbook2009/index.htm. Accessed 17 Sep 2010
30. World Resources Institute (WRI) in collaboration with United Nations Development Programme, United Nations Environment Programme, and World Bank (2008) World Resources 2008: roots of resilience—growing the wealth of the poor. WRI, Washington, DC
31. Veleva V, Ellenbecker M (2001) Indicators of sustainable production: framework and methodology. J Clean Prod 9:519–549
32. Neely A, Mills J, Platts K, Richards H (2000) Performance measurement system design: developing and testing a process-based approach. Int J Oper Prod Manage 20(10):1119–1132
33. International Organization for Standardization (ISO) (2006) ISO 14040: environmental management: life cycle assessment, principles and framework. http://www.iso.org. Accessed 16 Sept 2010
34. Ragas A, Knapen M, Heuvel P, Eijkenboom R, Buise C, Laar B (1995) Towards a sustainable indicator for production systems. J Clean Prod 3:123
35. Quinn M, Kriebel D, Geiser K, Moure-Eraso R (1998) Sustainable production: a proposed strategy for the work environment. Am J Ind Med 34:297–394
36. Griesshammer R, Buchert M, Gensch CO, Hochfeld C, Manhart A, Rüdenauer I, Ebinger F (2007) PROSA—product sustainability assessment. Meisterdruck, Freiburg

37. BASF (2010) SEEBALANCE®. http://www.basf.com/group/corporate/de/sustainability/eco-efficiency-analysis/seebalance. Accessed 15 Sep 2010
38. Schmidt WP, Taylor A (2006) Ford of Europe's product sustainability index. Proceedings of the 13th CIRP conference in life cycle engineering, Leuven, Belgium
39. Heijungs R, Huijbregts MAJ (2004) A review of approaches to treat uncertainty in LCA. Complexity and integrated resources management. Proceedings of the second biennial meeting of the International Environmental Modeling and Software Society (iEMSs)
40. Matthews HS, Hendrickson CT, Weber CL (2008) The importance of carbon footprint estimation boundaries. Environ Sci Technol Viewp 42(16):5839–5842
41. Suh S, Lenzen M, Treloar GJ, Hondo H, Horvath A, Huppes G, Jolliet O, Klann U, Krewitt W, Moriguchi Y, Munksgaard J, Norris G (2004) System boundary selection in life-cycle inventories using hybrid approaches. Environ Sci Technol 38(3):657–664
42. Lenzen M (2002) A guide for compiling inventories in hybrid life-cycle assessments: some Australian results. J Clean Prod 10(6):545–572
43. Suh S, Huppes G (2005) Methods for life cycle inventory of a product. J Clean Prod 13 (7):687–697
44. Bouman M, Heijungs R, Van der Voet E, Van den Bergh J, Huppes G (2000) Material flows and economic models: an analytical comparison of SFA, LCA, and partial equilibrium models. Ecol Econ 32:195–216
45. Cooper J (2000) Material flow analysis. University of Washington College of Engineering. http://faculty.washington.edu/cooperjs/Definitions/materials_flow_analysis.htm. Accessed on 2 Nov 2009
46. Cooper J, Fava J (2006) Life cycle assessment practitioner survey: summary of results. J Ind Ecol 10(4):12–14
47. Jolliet O, Margni M, Charles R, Humbert S, Payet J, Rebitzer G, Rosenbaum R (2003) IMPACT 2002: a new life cycle impact assessment methodology. Int J of Life Cycle Assess 8(6):324–330
48. Zhou X, Schoenung J (2004) Development of a hybrid environmental impact assessment model: a case study on computer displays. Proceedings of IEEE international symposium on electronics and the environment, Phoenix, AZ
49. Masanet E (2004) Environmental and economic take-back planning for plastics from end-of-life computers. Ph.D. dissertation, University of California, Berkeley
50. McKone TE (1999) The rise of exposure assessment among the risk sciences: an evaluation through case studies. Inhal Toxicol 11:611–622
51. Ramaswami A, Milfor JB, Small MJ (2005) Integrated environmental modeling: pollutant transport, fate, and risk in the environment. Wiley, Hoboken, NJ
52. National Research Council (NRC) (1983) Risk assessment in the Federal Government: managing the process. National Academy, Washington, DC
53. Brown CDJ (1997) Theoretical and mathematical foundations of human health risk analysis: biophysical theory of envronmental health science. Kluwer Academic, Boston, MA
54. Louvar JF, Louvar BD (1998) Health and environmental risk analysis: fundamentals with applications. Prentice Hall, Upper Saddle River, NJ
55. Kammen DM, Hassenzahl DM (1999) Should we risk it? Exploring environmental, health, and technological problem solving. Princeton University Press, Princeton, NJ
56. Paustenbach DJ (ed) (2002) Health risk assessment: theory and practice. Wiley, New York
57. McDaniels T, Small MJ (Eds). (2004) Risk analysis and society: interdisciplinary perspective. Cambridge University Press, Cambridge
58. Hertwich EG, Mateles SF, Pease WS, McKone TE (2001) Human toxicity potentials for life cycle assessment and toxics release inventory risk screening. Environ Toxicol Chem 20(4):928–939
59. Huijbregts MAJ, Thissen U, Guinée JB, Jager T, Kalf D, Meent D, Ragas AMJ, Sleeswijk AW, Reijnders L (2000) Priority assessment of toxic substances in life cycle assessment. Part I:

Calculation of toxicity potentials for 181 substances with the nested multi-media fate, exposure and effects model USES–LCA. Chemosphere 41(4):541–573

60. Jin X, High KA (2004) A new conceptual hierarchy for identifying environmental sustainability metrics. Environ Prog 23(4):291–301
61. Boyd S, Dornfeld D, Krishnan N (2006) Life cycle inventory of a CMOS chip. Electronics and the environment. Proceedings of the 2006 I.E. international symposium on electronics and the environment, San Francisco, CA, pp 253–257
62. Reich-Weiser C, Dornfeld DA (2008) Environmental decision making: supply-chain considerations. Trans North American Manufacturing Research Institute, pp 325–332
63. Superfund (2004) Comprehensive environmental response, compensation, and liability act. United States Environmental Protection Agency
64. Graedel TE, Allenby BR (2002) Hierarchical metrics for sustainability. Environ Qual Manage 12(2):21–30
65. Fleschutz T (2010) Contribution to sustainable industrial value creation by the multi attributive evaluation of assembly equipment. PhD thesis, TU Berlin, Fraunhofer Verlag, Stuttgart, Germany
66. Gausemeier J, Fink A, Dornfeld D, Ziemba B, Whitbeck R (2006) A decision-based analysis of compressed air usage patterns in automotive manufacturing. J Manuf Syst 25(4):293–300
67. Sudhoff W (2007) Methodology for the evaluation of locating overlapping mobility in production, PhD thesis, Technischen Universität München
68. Azapagic A (2003) Systems approach to corporate sustainability: A general management framework. Process Safety and Environmental Protection 81(5):303–316
69. Olsmats C, Dominic C (2003) Packaging scorecard—a packaging performance evaluation method. Packag Technol Sci 16:9–14
70. Schwartz J, Beloff B, Beaver E (2002) Use sustainability metrics to guide decision making. Chem Eng Prog: 58–63
71. Lapkin A (2006) Sustainability performance indicators. In: Dewulf J, Van Langenhove H (eds) Renewables-based technology: sustainability assessment. Wiley, New York
72. Tanzil D, Beloff B (2006) Assessing impacts: overview on sustainability indicators and metrics. Environ Qual Manage (Summer) 15(4):41–56
73. Seager TP, Satterstrom FK, Linkov I, Tuler SP, Kay R (2007) Typological review of environmental performance metrics (with illustrative examples for oil spill response). Integr Environ Assess Manage 3(3):310–321
74. Reich-Weiser C, Vijayaraghavan A, Dornfeld DA (2008b) Metrics for sustainable manufacturing. Proceedings of the 2008 international science and engineering conference, Evanston, IL

Green Supply Chain

4

Yifen Chen, Rachel Simon, Corinne Reich-Weiser,
and Justin Woo

*When one tugs at a single thing in nature, he finds it
attached to the rest of the world.*

John Muir

Abstract

This chapter discusses the background, characteristics, and requirements/
constraints of supply chains. A major focus is how they have been developing
towards more green performance. Several case studies are given documenting
the activities of companies that have taken leadership in green supply chain,
material reuse, and recycling. Problems associated with implementing green
supply chains are discussed.

4.1 Motivation and Introduction

As environmental awareness increases, concerns about green supply chains become
more pronounced in the business world. Traditionally, companies only focus on
their internal environmental measures of waste-reduction, pollution control, and
hazardous material control to comply with existing regulations and address their
social responsibilities. These days, companies have started to realize they have do
something more than these internal controls. One reason for the change is that
customers and other stakeholders do not distinguish between a company and its
suppliers [1]. Moreover, environmental "surprises" can cause financial harm
through disruptions in the supply chain or product liability [2]. Similar to product
recall, even if resulting from a manufacturing problem upstream in the supply

Y. Chen • R. Simon (✉) • C. Reich-Weiser • J. Woo
University of California at Berkeley, 1115 Etcheverry Hall, Berkeley, California 94720, USA
e-mail: rachelrific@gmail.com; corinnerw@gmail.com

D.A. Dornfeld (ed.), *Green Manufacturing: Fundamentals and Applications*,
DOI 10.1007/978-1-4419-6016-0_4, © Springer Science+Business Media New York 2013

chain, the public usually blames the brand owner for the defective product instead of the supplier who provides the defective component or inappropriate material content. These events often hurt the brand owner's reputation more than the small upstream supplier. Finally, as described in earlier, risks to business associated with supply chain externalities, such as government interference, drought, transport interruptions, and so on, can be minimized or avoided. Hence, greening the supply chain helps companies to avoid potential environmental problems that might threaten their overall environmental performance, reputation, and business success.

In the past decade, the trend toward green supply chains has not been clearly visible either in practice or in academic studies. In practice, with increasing energy and commodity costs, more and more companies started to extend their green house gas emission evaluation from direct emission (Scope 1^1) and indirect emission from purchased electricity (Scope 2) to the indirect emissions upstream and downstream of the supply chain (Scope 3). To help companies evaluate the environmental impact of their supply chains, one of the largest enterprise resource planning (ERP) software providers, SAP, announced the availability of a new solution, called SAP Environment, Health and Safety Management, in May, 2009 [4]. ERP software, which was introduced in the 1990s, integrates the information flow in a business and has helped many companies improve their supply chain efficiency. As it has an important role in supply chain management, the extension of this software to include sustainability shows the significance of insuring a green supply chain.

In academia, research on green supply chains has been increasing during the past decade (see, for example, Sarkis, [1]). The early researchers in this field focus on the reverse logistics of a closed-loop supply chain, such as recycling routing decisions, remanufacturing, disassembly, and testing. Now, researchers increasingly take a broader perspective to include considering the overall system effect. For example, Guide and Van Wassenhove [5] considered the remarketing of a reuse/remanufactured product and propose a framework to value the potential economic. Moreover, they shift the focus from cost minimization to value creation. Researchers in green supply chains also have been forced to be more interdisciplinary and the researchers in different fields, including economics, political science, and engineering, need to be coordinated. Currently, some researchers use economic models to analyze the effectiveness and efficiency of some legislation to have an impact on the "greenness" of a supply chain [6, 7] Fig. 4.1.

The remainder of this chapter is organized as follows. Section 4.2 provides a definition of a green supply chain. Section 4.3 discusses the issues and difficulties of implementing green supply chain strategies. Section 4.4 provides a brief overview of the methodologies and techniques applied to this area. Finally, Sect. 4.5 concludes the chapter and points out future directions of green supply chains.

---

[1] The GHG Protocol Corporate Standard defines three scopes of GHG emission and require companies to report their Scope 1 and Scope 2 GHG emission, While the Scope 3 emission is optional [3].

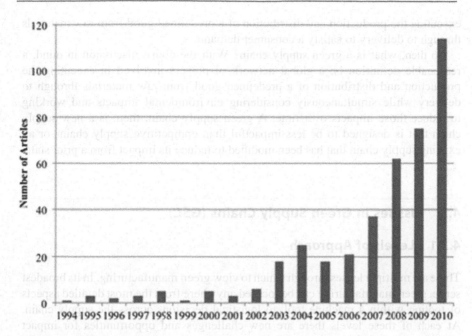

Fig. 4.1 Number of articles which are identified by a search of "green supply chain" OR "sustainability AND supply chain" in topic (search is performed on Web of Science on January 7, 2011)

4.2 Definition

While the aims of green supply chains are noble, and each element considered by researchers critically important for inclusion in a green supply chain, they are certainly not all-encompassing or holistic goals for a green supply chain designer to consider. As we will see throughout this book, studies that focus too narrowly on a single aspect or environmental consideration often fail to notice an unintended consequence of the suboptimization. It is no different for green supply chain research. A green-supply chain must, from design through execution, production through end of life, embody the goals of any sustainable manufacturing strategy: (1) appropriate goals, (2) comprehensive and repeatable metrics, (3) decision-making strategies. And then be followed by execution at the enterprise, system, machine, or process level using capable business or engineering tools.

Therefore, how do we define a "green supply chain?" First, what does it mean to be green? In this book, green is presented as different from the concept of sustainability. Where sustainability is a goal to leave adequate resources for future generations in terms of the environment, society, and the economy, the goal of being "green" is simply to reduce environmental impacts over time (without acknowledging when or if the level of impact is sustainable). A definition of a supply chain can be proposed as: the global network of players involved in executing the production and distribution of a predefined good from raw materials through to delivery to satisfy a consumer demand.

So then, what is a green supply chain? With the above discussion in mind, a reasonable definition is: a global network of players involved in executing the production and distribution of a predefined good from raw materials through to delivery while simultaneously considering environmental impacts and working to reduce those impacts over time. A green supply chain, then, is a new supply chain that is designed to be less impactful than competitive supply chains or an existing supply chain that has been modified to reduce its impact from a prior state.

4.3   Issues in Green Supply Chains (GSC)

4.3.1   Levels of Approach

There are multiple lenses through which to view green manufacturing. In its broadest sense, green manufacturing can be focused anywhere from the most detailed aspects of tooling within a machine tool, to logistical decisions across the supply chain. At each of these levels there are new challenges and opportunities for impact reduction. And at each of these levels there is a spectrum of control over impacts that decreases temporally. Specifically, for the supply chain, there is the loss of control that occurs once the supply chain layout is established, followed by choices of packaging and transportation mode, which can be modified only in small pieces rather than completely overhauled as might be necessary for significant reductions.

Greening the supply chain exists as a goal that is often considered separate from product design; however, there is a critical linkage between the two as seen in Fig. 4.2. From product design through to packaging there is a clear flow of information. However, this information flow is not always linear, and in many cases decisions at the very end of this chain can be considered in the initial product design. For example, the product may be designed for maximum shipping efficiency. Decisions about when to subassemble and complete final assemblies may be based on shipping efficiencies and minimized packaging. Transportation modes may similarly be a factor in product design and the consideration of packaging. Then, supplier locations may be dependent on the preferred transportation mode.

Product Design → Parts Specifications → Materials Specifications → Supplier Location → Supply Chain Layout → Transportation Modes → Packaging

Given the complexity and sophistication in the organization of manufacturing systems and processes, accurate environmental analysis requires a keen understanding of this organization. Manufacturing can be broken into "levels of study" across two orthogonal frameworks (Fig. 4.2). From the perspective of the organization of the system, we can consider manufacturing processes as being composed of four levels, from the level of the individual devices where unit processes take place, through to that of the enterprise, incorporating all the activities in the manufacturing system, including supply chain externalities. These four levels are as follows:

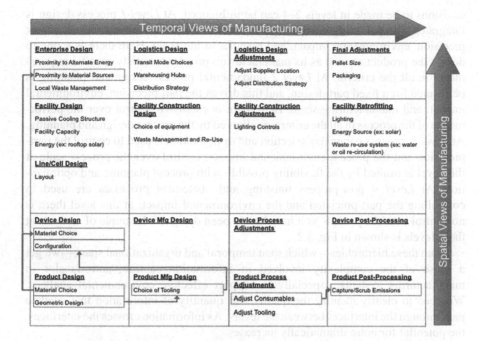

Fig. 4.2 Four temporal levels of manufacturing design—an integrated view of manufacturing design levels and the decisions they contain. *Arrows* represent flow of information from one decision to another [8]

1. Device—Individual device in the manufacturing system, which is performing a unit process. Support equipment for the unit process are included here such as gage systems and device-level oil-circulating systems.
2. Line/Cell—Logical organization of devices in the system that is acting in series or parallel to execute a specific activity (such as a part assembly). Support equipment for the collection of devices are included here, such as chip conveyers and tool cribs.
3. Facility—Distinct physical entity housing multiple devices, which may or may not be logically organized into lines, cells, etc. Support equipment required at the facility level are also included here, such as power generators, water purifiers, and HVAC systems.
4. Enterprise—The entire manufacturing enterprise, consisting of all the individual facilities, the infrastructure required to support the facilities, as well as the transportation and supply chain externalities.

An equally compelling—and orthogonal—view of manufacturing can be made through the design to manufacturing life cycle of the process being considered. Here, we start with the design of the product, and we work our way through the design of the manufacturing process, through to process optimization, and finally post-process control and abatement. These levels are temporal in nature, and indicate the degree of control over the environmental impact of the manufacturing process. Level 1 is the highest level and the earliest in design and manufacturing. At this stage all future decisions to be made in levels 2–4 can be influenced. At *Level 1* process design is integrated with part design and there is the most control over considerations of part precision, environmental impact, and manufacturing scale. Here there is scope to design the product as well as its manufacturing process to satisfy specific requirements in all the criteria. At *Level 2* fundamental process design and planning is performed for a fixed part design, and this drives the part precision, environmental impact, and manufacturing scale. Here there is extensive control over the performance of the process in all the criteria as allowed by the process design and planning. At *Level 3*, process parameter selection and optimization is used to control the part-precision and the process environmental impact; control over the process scale at this level is limited by the flexibility possible with process planning and optimization. At *Level 4* post-process finishing and abatement processes are used in controlling the part-precision and the environmental impact; at this level there is no control over the process as it has already been designed. Example of analysis at these levels is shown in Fig. 3.2.

From these hierarchies—which span temporal and organizational spans—we get a sense of the complexity involved in information capture, and transfer in manufacturing systems, especially to support effective environmental analysis. We need to clearly identify the quality and quantity of information that need to pass through the interfaces between the levels. As information crosses the interfaces, the potential for noise dramatically increases.

Stonyfield Farm Example

Stonyfield was originally established in 1983 as a nonprofit organic farming school for teaching sustainable agriculture practices [9]. Today, Stonyfield is the world's leading organic yogurt maker, selling over 200 million pounds of yogurt products in 2008 alone. Its products, which include yogurts, smoothies, soy-yogurts, and milk, are sold in leading supermarkets and natural food stores across the United States.

According to its mission statement, Stonyfield is committed to promoting the overall health of the planet through food, people, and business [4]. Specifically, Stonyfield has managed its supply chain through sustainable optimization of its transportation system. Stonyfield divides analysis of its truck transportation system into two parts: shipment of ingredients and shipment of finished products. By analyzing and optimizing its transportation operations, Stonyfield pinpointed areas for improvement in distribution issues, resulting in reduction of absolute carbon dioxide emissions by over 40% from 2006 to 2008 [9].

To optimize the incoming material transportation operations, Stonyfield purchases full truckloads when possible. In addition, Stonyfield backhauls, utilizing product delivery trucks by using them to also pick up ingredients in the same trip. For example, trucks delivering products to Boston, Massachusetts are also utilized to pick up organic fruit puree from suppliers based in Massachusetts.

(continued)

In analysis of transportation of its outgoing yogurt products, Stonyfield implemented an electronic system to gather transportation data. Utilizing the data, Stonyfield redesigned its outgoing shipment operations, effectively eliminating 95% of its less-than-truckload (LTL) shipments. Previously, LTL truckloads were shipped to central delivery terminals by region; each intermediate terminal had additional trucks to deliver product orders to customers. By removing its intermediate terminals and instead loading full truckloads, Stonyfield simultaneously enabled direct delivery to its customers and reduced the number of required trucks. Ultimately, this resulted in a greater efficiency of truck route utilization, reducing the number of miles traveled and harmful environmental output. Cost savings due to reduced fuel use from 2007 to 2008 are estimated to be US $2.5 million [9].

In 2009, Stonyfield began to utilize railway trains for cross-country shipment of a portion its products due to an EPA estimate that transportation via railway requires 11 times less energy than by truck [9]. Currently, Stonyfield is further pursuing efforts to decrease its transportation emissions via use of refrigerated train cars.

Interface Example

Interface, founded in 1973, is the world's largest modular carpet manufacturer, with sales in 110 countries and manufacturing facilities on four continents [10]. In 1994, Interface established its first ideas of approaches to environmental sustainability. Today, the company follows a "Mission Zero" goal of eliminating its negative environmental impact by 2020.

One technique by which Interface is striving to reach its "Mission Zero" goal is through its ReEntry Carpet Reclamation Program, which has prevented an estimated 81 million pounds of carpet waste being deposited into landfills since 1994 [11]. Each year, five billion pounds of waste and scrap carpet are dumped in landfills. The ReEntry 2.0 Program, which was revised and reestablished in 2007, recycles end-of-life carpet by separating carpet fibers from its backing. This process of clean separation, which was pioneered by Interface, enables recycling of Nylon 6,6 fiber, a carbon intensive carpet component [12].

Through this process of clean separation, end-of-life carpets are separated into nylon fibers and vinyl backings, both of which undergo further recycling treatment for reuse. Post-consumer Nylon 6 and Nylon 6,6 fiber is returned to the supplier, reprocessed, and recycled for use [13]. After combination with additional original materials, the recycled nylon fiber is used in new carpet fiber. The separated vinyl backing material is recycled via Interface's Cool Blue backing technology, which utilizes renewable resources for production of new carpet product. Together, the recycle nylon fiber and vinyl backing

(continued)

content is used to produce Interface's Convert carpet, which has an overall reduced embodied energy [11].

Xerox Example

Xerox manufactures and sells printers, digital production printing machines, and related supplies. Xerox has a Waste-Free goal "to create products that minimize waste during manufacturing, use by the customer, and at the end of products' life cycles" [14]. Xerox's goal in preventing and managing waste is "to design products, packaging, and supplies that make efficient use of resources, minimize waste, reuse material where feasible and recycle what can't be reused." [15]. It does this through various programs and means, one of which is its take-back and recycling program. Xerox first began remanufacturing and recycling techniques in 1991. In 2003 alone, 161 million pounds of waste was prevented from being placed in landfills as a result of equipment remanufacture, reuse, and recycle [14]. Financially, remanu-facturing equipment results in a several hundred million dollar savings per year for Xerox [14]. Today, Xerox continues the practice and has since prevented over two billion pounds of waste from being dumped in landfills [15]. In 1991, Xerox first pioneered the practice and concept of remanu-facturing parts from its office equipment. Xerox's machines "are designed for easy disassembly and durability, contain fewer parts, and are controlled for chemical content," [15], all of which enable an efficient and beneficial remanufacturing process.

Xerox provides its customers with various options to recycle and dispose of equipment through Xerox-run collection sites or pick-up services. Xerox's Green World Alliance Program is one implementation designed to further enhance their Waste-Free strategy; through the Green World Alliance, printer cartridges, toner casings, and left-over toner can all be recycled through a closed-loop recycling system within Xerox [14]. Incentive and motivation for recycling is provided through a service that provides prepaid postage and packaging for customers to ship end-of-life equipment or materials for recycling; new supplies, such as toner cartridges, already include return labels in the product boxes (Xerox [14]). Leased products are picked up from customer sites for recycling by Xerox upon renewal terms, while sold products are picked up upon equipment replacement. For customers not continuing such leasing services, Xerox provides a free recycling service; customers pay shipping costs, but Xerox covers the cost of recycling [16]. Xerox collects its end-of-life equipment through these programs and reuses 70–90% (by weight) of each machine's components. This does not sacrifice standards, however, as performance specification requirements are still met and machines using remanufactured parts function as new machines. Only

(continued)

parts that meet all reliability and functional standards for the life of the product are reused. Xerox ensures that all products, regardless of the amount of reprocessed material or content, meet the same quality and reliability performance standards [14].

By taking into account remanufacturing, recycling, and reuse into the design process, Xerox further enhances the success of the program. Reprocessing parts is an actual design issue that is taken into account in the design process, before the equipment is actually made. Thus, Xerox is able to increase end-of-life potential for its products. Today, 100% of its newly-designed equipment takes remanufacturing into consideration during the design phase [14]. In designing products, Xerox utilizes a modular product architecture that uses a core set of components that makes the remanufacturing and reuse process more efficient. With options for reuse more versatile due to the same core design setup, Xerox can use similar parts from one machine type in a different machine type, or even for future generations of the same model. This results in models having up to 60% of parts by weight in common with previous equipment [17]. Ultimately, this results in overall reduction of energy consumption, raw material use, equipment cost, and environmental impacts.

4.3.2 General Problem Faced in GSC

In those issues of green supply chain mentioned above, each of them has different difficulty to achieve. Still, there are some general problems in the practice of green supply chain.

First of all, to implement green strategies, a company has to decide whether it should change to another greener supplier or stick to the original suppliers and work with them toward sustainability. Simpson and Power [18] provided a major review of the green supplier partnership and concluded that transaction costs and efficacy of approach for the buyer are the critical issues in improving or influence a supplier's environmental performance. In supply chain management, it is well known that maintaining good partnership with suppliers is one of the key success factors. With a long-term partnership, the company could negotiate the price and quality and adjust the delivery time easier in an emergency. However, working with current suppliers to achieve sustainability may take more time and cost than switching to another supplier if current suppliers don't have the capability or equipment toward sustainability. On the other hand, switching to a new supplier will incur some intangible switching cost, such as time to select a greener supplier and effort to reconfigure.

The best strategy depends on the current environment of the company and the industry. However, there are some general rules for making a good decision. First of all, a company has to scrutinize whether the components from the suppliers in question are essential in the core competitiveness of the company or not. If the component or service is essential and complex, it might be better for them to stick on the original suppliers. However, if this component or service is only a supported function and the cost of changing suppliers is relative low, this company might choose a new supplier. For example, when HSBC try to decrease the carbon footprint of its corporation travel, they choose their suppliers by picking those that have the same methodologies and principles as they do. Other companies require candidate suppliers meet some predefined criteria and continuously work with them toward sustainability once the contract is signed. For example, Levi Strauss selects the suppliers based on their Terms of Engagement and continuously works with their supplier by monitoring and assessing their suppliers and identifying areas of improvement.

On the other hand, Levi Strauss chooses to work with existing suppliers toward sustainability.

Levi Strauss Example

Levi Strauss was founded in 1873 and today is one of the world's largest brand-name apparel companies, selling jeans and casual pants in over 110 countries [19]. Its supply chain is expansive, including over 600 external suppliers over a network of over 50 countries and 350 facilities [20]. In 1991, Levi Strauss made history when it became the first major multinational company to establish Global Sourcing and Operating Guidelines, which held contractors along its supply chain accountable to certain sustainability standards [21].

Today, the Global Sourcing and Operating Guidelines is core to Levi Strauss' business partnerships along its supply chain. Each business is evaluated according to explicit Terms of Engagement listed in the Global Sourcing and Operating Guidelines; some points of evaluation include environmental requirements, employment standards, community involvement, and employment standards [22].

\Prior to establishing a partnership relationship with a prospective factory or supplier, Levi Strauss reviews the Terms of Engagement (TOE) guidelines, which include factory and supplier-specific expectations for the business. All suppliers are required to sign a document indicating intention of compliance with Levi Strauss' TOE expectations and corrective action plans [23]. All new and existing supplier partnerships are inspected annually by 20 full-time factory assessor Levi Strauss employees. Annually, the factory assessors audit each supplier to review its performance with regards to Levi Strauss' Terms of Engagement guidelines. Factory assessors investigate and evaluate supplier practices by four methods: (1) interviews with facility management, (2) review of working and wage hour records, (3) physical inspections of factory health and safety conditions, and (4) discussions with factory workers [23]. Using these four methods of inspection, assessors are able

(continued)

to evaluate social, labor, and environmental sustainability for a given supplier. Improvement areas and corrective actions are recommended to the suppliers based upon a three-category system of zero-tolerance, immediate-action, and continuous-improvement items; these are subsequently reported by the assessors to the Vice President for Global Code of Conduct [23].

In combination with the Terms of Engagement, Levi Strauss developed an assessment tool to identify areas of improvement for supplier factories. Levi Strauss' Supplier Ownership Guidebook details a management system approach and its application to the Global Sourcing and Operating Guidelines; in addition, Levi Strauss holds a two-day code of conduct training program for factory executives [23]. The purpose of such initiatives and programs is to promote supplier ownership over the Global Sourcing and Operating Guidelines, ultimately expanding sustainability and the knowledge of its importance.

In addition, Levi Strauss has established Global Effluent Guidelines for all suppliers, expanding its Global Sourcing and Operating Guidelines for specific application to environmental sustainability. According to life cycle assessments, 49% of Levi Strauss' water consumption occurs during its cotton production phase for clothing products [20]. Accordingly, Levi Strauss' Global Effluent Guidelines mandates strict wastewater contaminant levels for its suppliers worldwide [20]. In 2008, Levi Strauss partnered with 13 other companies to form the Sustainable Water Group; together, this group extended Levi Strauss' Global Effluent Guidelines to all second-tier suppliers of bulk fabrics and sundry items [20]. Through promotion of these guidelines, Levi Strauss promotes effective environmental sustainability at multiple levels of its supply chain through water management processes.

In addition, determining the most sustainable option may not always be a straightforward task. Assessments of sustainability depend upon which metrics are being considered and on what scale. Munda [24] has demonstrated that the rankings of sustainability assessments can change based on which metrics are chosen. Also assessments can vary dramatically depending on which aspects of a product's lifecycle and supply chain are considered. Bearing this in mind, the manufacturing and geographical scope of assessments must clearly be identified [25].

Even when metrics have been generally agreed upon, sustainability decisions can be difficult to make. In many instances, any improvements in the supply chain are obtained at a tradeoff with another sustainability or business impact. For instance, Reich-Weiser and Dornfeld [26] show that greenhouse gas emissions inventories may be altered when the supplier's location is changed. In this case, the three location-based factors that contribute to emissions are: transportation distance; mode of transportation; and the type of technologies used for electricity generation. Tradeoffs between these different sources of emissions can be achieved by changing production facilities.

Moreover, generally recognized green conventions and best practices do not necessarily provide improvements in sustainability impacts. At present there exist several prescribed methods for greening one's product or operations. However, when implementing such practices, great care must be taken to ensure all factors across the supply chain have been taken into consideration. For instance, many packaging suppliers have adopted the practice of lightweighting, or making components that weigh less. This is done by using less material, or a material that has a lower density. Lighter products can often lead to reductions in resource use, waste creation, and environmental impacts related to transport and raw material extraction. However, there are some instances in which lightweighting may not be sustainably beneficial. For instance, if a reduction in materials results in an increase in the rate of damage during shipment, then the benefits of a lighter product could be negated. In this case, the net effects on environmental impacts could be worse than it was prior to the implementation of lightweighting.

On the other hand, some consider it to be more sustainable to design products that are more durable. This can reduce the frequency with which a product must be replaced. As a result the overall long-term resources used and waste created can be lessened. Increased durability is often achieved by using more materials or increasing the complexity of manufacturing involved. A simple example of this is a cloth or durable plastic tote, which is used in place of paper or plastic bags provided by retailers at the point of purchase. Much like in lightweighting, making products more durable can have sustainability effects that are the opposite of what was intended. This can happen if consumers replace or dispose of the product before the end of its useful life, resulting in an impact that is worse than the original alternative.

Additionally, other factors outside of the manufacturer's control can affect a producer's sustainability impacts. For example, policy and industry practices play a part in the amount of infrastructure that is available for post consumption recovery. Furthermore the scale upon which a practice has been adopted by industry affects the availability and cost of a particular choice.

Patagonia Example

One company that has been challenged by this situation is Patagonia. They faced this predicament in 1996, with their decision to use cotton that is only organically sourced. This decision grew out of a determination to investigate the company's environmental impacts. During this process employees met an organic farmer and activist, who showed them of the impacts of conventional cotton farming. While on farm tours, they were taken aback by the noticeable difference between organic and conventional agricultural practices. This is because conventional farming has many side effects that impact the quality of the land, many of which are readily apparent. For instance, over time conventional cotton production depletes soil of its fertility. As a result, farmers must use increasing amounts of synthetic fertilizers just to maintain yields. In addition, pesticides are often required—applied in an aerial

(continued)

fashion—to keep insects away. In aggregate these supplementary additives can severely pollute waterways, and permanently affect the productivity of a farm and its surrounding land. Patagonia representatives were deeply affected by the stark visual contrast between conventional and organic farming practices.

Patagonia faced many obstacles in transitioning to organic cotton. They had difficulty getting their suppliers to comply with their request. The entirety of their textiles supply chain was inexperienced with organic materials. Fabric vendors had difficulty finding organic finished fabrics. Dying mills did not know producers of raw organic fabrics. Knitters and weavers did not know where to get organic yarn. Yarn spinners did not have organic sources. Further, the quantities that Patagonia wanted were small, making it more difficult for these vendors to accommodate such a request. Eventually Patagonia was able to find organic suppliers. Though, this was only possible through a great effort on their part to create linkages across the supply chain. In addition, they were only able to find suppliers in the United States, far from their existing spinning mills. Even at present day, their options for sourcing locations remain limited. To illustrate, in 2007, organic cotton production was concentrated to the countries of: India with 51% of global production; Syria with 19% of global production; and Turkey with 17% of global production [27]. In addition China, Tanzania, and the US were also options for organic cotton supplies, as they contributed 5%, 2%, and 2%, respectively, to the world's global stocks in 2007.

All in all, the results of Patagonia's efforts were mixed. Given the company's commitment to adopting organic cotton, they were successful in achieving this immediate goal. However, they had also hoped that their efforts would influence others in the apparel industry. They were less successful in this objective. As of 2006, 10 years after Patagonia's switch, organic cotton only accounted for 0.1% of the global retail apparel industry.[2]

Additionally, many researchers measure environmental impacts based solely on green house gas emissions. In this respect it is questionable whether Patagonia's switch to organic had a positive impact. It should be noted that for organic versus nonorganic cotton production, this type of assessment is difficult to estimate, due to the differences in climates and farming practices around the world. Despite the complexities involved, Budden [30] notes that from various estimates the range of emissions from conventional cotton grown in the US can be as much as three times as high as the emissions

(continued)

---

[2] Cygnus Business Consulting and Research [28] estimates the 2006 global apparel retail industry to be valued at $852.8 billion. According to SBI [29] organic cotton retail sales in 2006 were $1.1 billion. Of this amount, they estimate 85% to account for apparel, with the remaining 15% accounting for home textiles. Using these figures, organic cotton apparel retail sales for 2006 are valued at $1.275 billion, or 0.1% of all retail apparel.

from organic cotton (6 kg of CO_2 equivalent per kg of fiber for conventional versus 2 kg/ kg for organic). Using these figures, raw cotton for a T-shirt from Patagonia's fall 2009 line would result in 1.2 kg of CO_2 equivalent emissions for conventional cotton fiber and 0.4 kg for organic cotton fiber.[3] Meanwhile, Patagonia [31] estimates that the shipping and facility energy use emissions are 1.6 kg, which is greater than the impacts of either cotton production choice. Similarly, for a polo shirt from their fall 2009 line, conventional cotton production emissions are approximately 1.57 kg, and organic cotton production emissions would be 0.52 kg.[4] Shipping and facilities energy use emissions are given as 12 kg of CO_2 equivalent [31]. In both of these cases the impacts of cotton sourcing are less than the emissions from transportation and facility energy use. Beyond this, many LCAs have shown that for garments, the energy use or global warming impacts from the use phase (i.e., the impacts of laundering) are far greater than the impacts from any other part of the life cycle [32–35]. For instance, Levis Strauss estimates that laundering a garment 24 times (once every 2 weeks for a year) results in 16.9 kg of CO_2 equivalent emission [36].

Despite the complexities involved in sustainable decision making, producers' efforts to improve impacts along their supply chains are generally beneficial. At the very least, such attempts increase people's awareness of the issues involved. They also set a precedent for others within their industry. Supply chains are complex, and attempts by companies to improve their impacts results in increased transparency across the product life cycle. Even an imprudent decision can advance sustainability efforts, as it incites discussions about the impacts of different alternatives. Any such efforts also contribute to the body of sustainability assessment work, as researchers support or refute a company's claims of improved sustainability. Despite any missteps, improvements are generally brought to fruition through a producer's strong commitment to melioration. Initial improvements are made by taking advantage of inefficiencies in operations. As companies exhaust these opportunities, subsequent goals require increasing effort to achieve.

---

[3] Patagonia [31] estimates the CO_2 emissions (from production and transport) associated with their T-shirt to be 1.6 kg. They also estimate the weight of the shirt to be one-eighth the weight of this figure, which can be calculated as 0.2 kg. Using Budden's [30] estimates for the impacts of cotton production, conventional cotton fiber production would be six times the weight, or 1.2 kg, and organic cotton production would be two times the weight, or 0.4 kg.

[4] Patagonia [31] estimates the CO_2 emissions (from production and transport) associated with their polo shirt to be 12 kg. They also estimate that these emissions are 46 times the weight of the shirt. The weight of the shirt can then be calculated as approximately 0.26 kg. Using Budden's [30] estimates for the impacts of cotton production, conventional cotton fiber production would be six times the weight, or 1.57 kg, and organic cotton production would be two times the weight, or 0.52 kg.

The potential for a company to make a positive impact is especially great when the company holds a leadership position within their industry. Many manufacturers use third party suppliers for various aspects of their operations, especially for segments that are not a part of their core business. Meanwhile, suppliers often provide parts and materials for multiple parties within an industry. In many instances, it is not in the interest of a supplier to create products that are substantially different from those demanded by their primary business partner. So if a supplier's largest client changes their requirements, it may alter what the supplier is able to provide to their other customers (i.e., to the competitors of their largest client). Even when suppliers are solely dedicated to one company, product modifications from their clients can affect others within the industry. As the scale of demand for particular components or raw materials changes, prices can be affected. In some industries, these changes may cause an overall shift in practices that follows in the footsteps of the manufacturer with the largest market shares.

One such instance of this is when McDonalds decided to end their use of polystyrene packaging. In 1989 McDonalds was at the center of criticism for their use of polystyrene clamshell packaging. Polystyrene had been specifically identified by the public as representative of society's waste issues. As the largest global fast food chain, McDonalds had been identified as a major contributor to the problem. Despite the fact that McDonalds very minimally used polystyrene and actively made efforts to recycle it [37], their influence and visibility made them the target of anti-polystyrene campaigns. The efforts against McDonalds involved demonstrations, letter writing, and consumer mailings of clamshells back to the company. At the same time, many cities within the US were considering polystyrene bans. All paper and plastic foodservice packaging for the company was handled by one privately owned company, Perseco. Perseco worked with numerous companies helping them meet their packaging needs. They also procured goods from over one hundred different suppliers, several of which provided polystyrene [37]. So, when McDonalds finally made the decision in 1990 to stop using the material, they affected numerous producers of polystyrene products. McDonalds affected the industry by changing the overall demand for polystyrene and making the debate over the environmental impacts of the plastic highly visible. The resulting uneasiness felt by the polystyrene industry was evidenced by one supplier, Amoco Chemical, running a full-page advertisement in Business Week claiming that bans and substitutions of polystyrene would not help "solve the problem."

Nau Example
Nau is an American outdoor clothing start-up company created from scratch with environmental and social sustainability goals and standards as part of its core founding mission. Unlike most other existing companies working towards sustainability through incentive programs or focus redesign, Nau approached sustainability from the ground up. In the workplace, sustainability was an organizational design constraint; all of its outdoor clothing

(continued)

products had to be sustainable or else its mission statement would not be upheld.

As a core part of its company mission, Nau established its "Rules for Corporate Responsibility," establishing hard guidelines by which the company would operate both internally and externally along its supply chain. One of the details from the Rules for Corporate Responsibility included donating 5% of the purchase price to community partners working towards sustainable change.

Secondly, Nau limited its colors for clothing to earth tones and black due to the need of harmful substances required to make brighter colors; this is a direct of example of sustainable policy on design. Following the strict environmental rules established in its Rules for Corporate Responsibility, Nau engineered 27 new fabric types for use in its clothing lines; in total, 27 out of the 30 fabric types used by Nau in its clothing were new [38].

The strong environmental mission of Nau was clearly stated in its Rules for Corporate Responsibility. However, these very rules made it difficult to gain sufficient funding from investors; in one case, investors even agreed to provide $5 million to Nau during its start-up stages under the condition that the Rules for Corporate Responsibility were removed. Balancing such environmental and business policy choices, as well as its subsequent effect on investors proved difficult. In his article "What Nau?" writer Luke O'Brien writes, "The irony, of course, is that a company so committed to sustainability was ultimately unsustainable. This was not because of its principles, but rather the mundane problems that plague most start-ups—lack of money and poor execution—had undermined Nau" [38].

Founded in 2005, Nau had raised $34 million of capital by 2007 [39]; however, its 2007 sales were less than $6 million [39]. Thus, due to insufficient funds, in May 2008, Nau was forced to cease operations. A year later, Nau was bought by Horny Toad Activewear, a California-based clothing company [38]. Under new ownership, Nau underwent operational changes, such as a reduction in the donation amount from 5 to 2% and reduced employee workforce; its core mission is still commitment to environmental sustainability. Under new management, Nau has undergone a more streamlined process, with 75% of sales coming from online purchases [40].

4.4   Techniques/Methods of Green Supply Chain

In order to achieve the sustainability in a supply chain, many techniques and methodologies are using to measure, evaluate and compare the green performance of a supply chain. Srivastava [41] surveyed a complete literature review of green supply chain and categorized the issues in green supply chain management into green design and green operation. In this section, we give a brief review of those useful techniques and methodology in green supply chain, including life cycle assessment, optimization programming, statistical analysis and game theory.

Life cycle assessment (LCA) is a technique tool that can identify the environmental impact associated with a product/process/activity and evaluated opportunities to reduce these impacts. The Society of Environmental Toxicology and Chemistry defines LCA as looking holistically at the environmental consequences associated with the cradle-to-grave life cycle of a process or product. LCA is a critical technique in greening supply chain since it evaluates the whole life cycle impact of a product across a supply chain. However, other techniques have to be combined with LCA to better evaluate the overall supply chain. Bevilacqua et al. [42] combined LCA and quality function deployment (QFD) to define design specifications for all the stakeholders involved in a supply chain: clients, manufacturers, suppliers, suppliers of the suppliers, etc. Hagelaar and Vorst [43] argued that different LCA techniques should be use when dealing with different supply chain structure. Chapter 3 covered life cycle analysis techniques in some detail.

Optimization programming is one of the well-known methodologies which is widely used in supply chain analysis and management research, which is also a good tool to evaluate the trade-off between different aspects such as environmental impact, production cost, and product quality. There are plenty of research papers that use optimization modeling while addressing varied issues in green supply chains, such as waste management, reverse logistics, and production planning under remanufacturing. For example, Richter and Dobos [44] applied integer nonlinear programming on economic order quantity (EOQ) repair and waste disposal model and find the optimal strategy which minimize the cost of a company. Dynamic programming is also commonly used when dealing with stochastic random variable, such as demand and inventory in different period. For example, Inderfurth et al. [45] use a period review model to find the optimal allocation policy for returned products under uncertain demand and return quantity.

Another prevailing methodology is statistical models, which are often used when empirical data is available to investigate the relationship between several factors. The result from statistical analysis allows development of a model for approaching issues of related fields. Chinander [46] interviewed the top and senior managers in a steel manufacturer and use ANOVA and other statistical test to assess the internal driver of environmental awareness in a firm. Haas and Murphy [47] used regression to compensate for nonhomogeneity of data envelopment analysis and compared models using actual data of municipal reverse logistic channel.

Game theory is another technique on supply chain which became popular in the past decade. It is usually applied to researches which study the interaction between several stakeholders in a supply chain or the efficiency of the mechanisms of government policies. Majumder and Groenevelt [48] analyzed a two-period competition game between one original equipment manufacturer (OEM) and one local remanufacturer. Atasu et al. [7] discussed the economic and environmental impacts of extended producer responsibility type of legislation and identified efficiency condition.

Besides the above mathematical technique, computer programming and commercial software are often applied to researches for practical purpose. Some researches offer approximated solutions that could be easier to compute so that these methodologies could be adopted in practice. Some research provides numerical analysis or case study to support the theoretical results in their researches. Kiesmüller and Scherer [49] demonstrated how to compute the optimal policy for a stochastic one product recovery system. They coded a simulation program to do the numerical analysis and suggested a high-quality approximation which can be computed quickly. McLeod and Cherrett [50] use an off-the-shelf vehicle routing software to aid the calculation of distance and then evaluate the environmental impacts for different routing option for small parcel transactions.

With these techniques and methodologies, the researchers have developed fundamental theories of the operation level of green supply chain. However, most theoretical research focuses on a single dimension of improvement. To investigate the holistic environmental impact of products and processes, the interdisciplinary nature of this field requires researchers to incorporate methodologies in different fields. Corbett and Klassen [2] suggested that the researches in green supply chain needs to expend the boundaries and integrate the concerns of more stakeholders, including governments, local communities, public interest groups, and future generations.

4.5   Future of Green Supply Chain

Despite all the efforts being made in greening the supply chain in the past few years, there are still several challenges that need to overcome. One of the fundamental problems is how do we define and measure environmental performance of a supply chain. Since the environmental problems of one supply chain may affect another one, the issue becomes complicated and may depend on the industry sector in which the company belongs. There are some measurements developed, such as balanced scorecard and cost-and-benefit analysis. However, there is no widely accepted standard existing in most industries and identifying, monitoring and acting on specific measures remains a challenge. For detailed discussion of metrics, readers are referred to Chap. 3 of this book.

Another big challenge is that collecting data from upstream suppliers or downstream retailers is not easy especially in some industries that have long supply chains. Current practices show that most big companies can only collect data from their first-tier suppliers to evaluate the environmental impact of a certain product.Even the largest retailer in the world—Wal-Mart—may have challenges obtaining data from their suppliers on their environmental impact. However, because of the size of Wal-Mart, the sustainability program initiated by them will have a significant influence.

Walmart Example

Walmart is an American-founded corporation that runs a chain of discount department stores; in 2010, it was the world's largest public corporation by revenue [51]. Thus, Walmart's supply chain policy decisions have a large influence upon suppliers along every tier of its supply chain. According to its own estimations, 92% Walmart's emissions come from its supply chain, with only the remaining 8% coming from their own operations [52].

To enhance transparency with suppliers, Walmart utilizes a Packaging Scorecard to measure the environmental sustainability of each supplier's packaging. This metric is an example of greater use of transparency, as it encourages suppliers not only to measure their emissions, but also to reduce them.

In 2009, Walmart introduced implementation of a Sustainable Product Index, a method by which to evaluate the sustainability of its supplier's products. Walmart promoted the development of a Sustainable Product Index to provide its customers with a more transparent supply chain; with more information about how a product's lifecycle and how it was made, customers will be able to make informed decisions when purchasing products. The new Index was rolled out in three phases and was in addition to the Packaging Scorecard metric. In the first phase of development of its Sustainable Product Index, Walmart provided its suppliers with a guide detailing both the purpose of the Index and the importance of being sustainable. The first phase also included a survey of 15 questions covering four major topics of sustainability including: (1) energy and climate, (2) material efficiency, (3) nature and resources, and (4) people and community. Walmart scored the suppliers' responses to each of the four major question topics separately, providing specific definitions and examples of its three possible score categories: "Below Target," "On Target," and "Above Target." Each supplier question response was first assigned a percentage value, which was in turn classified into one of the three category scores (Below, On, or Above Target). As with the Packaging Scorecard, suppliers were not required to complete the 15-question survey of the first phase; however, they were highly encouraged to do so [53].

The second phase included development of a lifecycle analysis database to compile information on products' lifecycles [5]. Walmart provided initial funding for a Sustainability Index Consortium, which is currently administered by Arizona State University and the University of Arkansas. The consortium, which consists of universities, suppliers, retailers, and governmental organizations, is intended to collaboratively develop a database of information with products' lifecycle information [53].

Lastly, the third phase, which is still under development, is to provide customers with product the Sustainability Product Index information; the optimal way of presenting easily accessible information has yet to be determined [53].

(continued)

As a large international company, Walmart also has defined several goals and standards for its suppliers in China. By 2012, Walmart's top 200 suppliers in China must become 20% more energy-efficient; this includes any and all areas of energy use and emissions, such as coal consumption, water use, energy use, and manufacturing processes. As motivation for Chinese suppliers, Walmart has stated that by 2012, they will cease orders from suppliers who do not meet the sustainability demands defined [54, 55]. The ultimate goal is to produce a win–win situation, with Walmart improving the environment, decreasing waste costs, and subsequently providing better service to its customership [54].

In addition, globalization makes it more difficult to cooperate with all the members in a supply chain. Most suppliers and manufacturers in developing countries have difficulty to improve the process or materials they use to become greener. One of the examples is IKEA's wood supplier. IKEA started its sustainability program in 1999. They stopped using plastic bags and invested $77 million in clean technology start-ups like solar. IKEA's supply chain efforts have helped thousands of cotton producers slash water and pesticide use. However the furnishings giant is struggling to source sustainable timber [56].

According to ENDS Report 2009, in 2006, IKEA promised 30% of the seven million cubic meters of solid wood used in its products annually would come from forests certified as responsibly managed by 2009. Its 2008 sustainability report, published in July 2009, admits this target will not be reached and shows only 7% of timber was certified in 2007/2008. IKEA said third-party verified timber is still in short supply in countries where its suppliers are based. Less than 10% of the world's forests are covered by a certification scheme and most of these are in the West [56]. This is mainly because in developing countries, most timber suppliers are unable or unwilling to manage forests in a sustainable way and pay the cost of certifications. Therefore, although the global supply chain provides the cost-saving benefit for a company so that they can find low-cost materials and components, the different cultures and economies of the countries where their suppliers are based increases the challenges of sustainable improvement over the supply chain.

References

1. Sarkis J (2006) Greening the supply chain. Springer, Berlin
2. Corbett CJ, Klassen RD (2000) Extending the horizons: environmental excellence as key to improving operations. Manuf Serv Oper Manag 8(1):5–22

3. World Resources Institute and World Business Council for Sustainable Development (2004) The greenhouse gas protocol: a corporate accounting and reporting standard (revised edition). Available at http://www.ghgprotocol.org/standards/corporate-standard. Accessed 10 Aug 2009
4. Kolakowski N (2002) sap applications focus on sustainability, Green IT. http://www.eweek.com/c/a/Enterprise-Applications/SAP-Applications-Focuses-on-Sustainability-Green-IT/. Accessed 23 Dec 2009
5. Guide VDR, Van Wassenhove LN (2001) Managing product returns for remanufacturing. Prod Oper Manag 10(2):142–155
6. Subramanian R, Gupta S, Talbot FB (2008) Product design and supply chain coordination under extended producer responsibility. Prod Oper Manag 18(3):259–277
7. Atasu A, Van Wassenhove LN, Sarvary M (2009) Efficient take-back legislation. Prod Oper Manag 18(3):243–258
8. Reich-Weiser C, Vijayaraghavan A, Dornfeld D (2010) Appropriate use of green manufacturing frameworks. Proceedings of the CIRP life cycle engineering conference, China, May 2010
9. Stonyfield (2011) About us. Company website. http://www.stonyfield.com/about_us/index.jsp. Accessed 4 Jan 2011
10. Interface (2011a) Interface's history. Company website http://www.interfaceglobal.com/Company/History.aspx. Accessed 5 Jan 2011
11. Interface (2005) Kicking the oil habit. http://www.interfaceflooring.com/library/PDF/CoolBlue.pdf. Accessed 5 Jan 2011
12. Interface (2011b) Closing the loop: ReEntry 2.0. Company website. http://www.interfaceglobal.com/Sustainability/Sustainability-in-Action/Closing-the-Loop.aspx. Accessed 5 Jan 2011
13. Interface (2011c) ReEntry 2.0 challenges what people believe is possible. Company website. http://www.interfaceflor.com/reentry2.0/. Accessed 5 Jan 2011
14. Xerox (2005) Environment, health and safety: 2005 progress report. http://www.bytes.co.uk/xerox/pdf/ehs_2005_progress_report.pdf. Accessed 31 Jan 2011
15. Xerox (2009) Prevent and manage waste. Company Website. http://www.xerox.com/about-xerox/environment/recycling/enus.html. Accessed 7 Sept 2009.
16. Xerox (2010) Environment, health and safety: xerox machine recycle/disposal methods. http://www.xerox.com/downloads/usa/en/e/ehs_recycle_options.pdf. Accessed 28 Dec 2010
17. Xerox (2008) 2008 report on global citizenship. http://www.xerox.com/Static_HTML/citizenshipreport/2008/citizenshipreport08.pdf. Accessed 31 Dec 2010
18. Simpson DF, Power DJ (2005) Use the supply relationship to develop lean and green suppliers. Supply Chain Manag 10(1):60–68
19. Levi Strauss & Co. (2011) Heritage. Company Website. http://www.levistrauss.com/about/heritage. Accessed 5 Jan 2011
20. Levi Strauss & Co. (2011) Sustainability. Company Website. http://www.levistrauss.com/sustainability. Accessed 5 Jan 2011
21. Levi Strauss & Co. (2011) Product Suppliers. Company Website. http://www.levistrauss.com/sustainability/product/product-suppliers. Accessed 5 Jan 2011
22. Levi Strauss & Co. (2011) Global sourcing & operating guidelines. Company Website. http://actrav.itcilo.org/actrav-english/telearn/global/ilo/code/levi.htm. Accessed 5 Jan 2011
23. Levi Strauss & Co. (2011) Levi Strauss & Co. Global sourcing and operating guidelines. http://www.levistrauss.com/library/levi-strauss-co-global-sourcing-and-operating-guidelines-0. Accessed 5 Jan 2011
24. Munda G (2005) Multi-criteria decision analysis and sustainable development. In: Figueira J, Greco S, Ehrgott M (eds) Multiple-criteria decision analysis: state of the art Surveys. Springer International Series in Operations Research and Management Science, Springer, New York
25. Reich-Weiser C, Vijayaraghavan A, Dornfeld DA (2008) Metrics for sustainable manufacturing. Proceedings of the 2008 international science and engineering conference, Evanston, IL

26. Reich-Weiser C, Dornfeld DA (2009) A discussion of greenhouse gas emission tradeoffs and water scarcity within the supply chain. J Manuf Syst. doi:10.1016/j.jmsy.2009.04.002

27. Organic Exchange (2008) Organic cotton farm and fiber report 2008. Organic Exchange, Berkeley

28. Cygnus Business Consulting and Research (2004) Executive summary of apparel retailing in India. http://www.cygnusindia.com/Industry%20Insight-Apparel%20Retailing%20in%20India-Executive%20Summary%20&%20TOC-March%202004_.pdf. Accessed 5 Jan 2011

29. SBI Reports (2008) The U.S. market for organic and eco-friendly home textiles. http://www.sbireports.com/Organic-Eco-Friendly-1495188/. Accessed 11 Oct 2010

30. Budden R (2007) The carbon challenge for textile manufacturers: the influence of climate change on choices between fibres and suppliers. International Textile Manufacturers Federation. http://www.textile.fr/sitepj/Budden.pdf. Accessed 4 Oct 2010

31. Patagonia (2009) The footprint chronicles. Company Website. http://www.patagonia.com/web/eu/footprint/index.jsp. Accessed 5 August 2009

32. American Fiber Manufacturers Association (1993) Life cycle analysis (LCA): woman's knit polyester blouse. Franklin Associates LTD. http://www.fibersource.com/f-tutor/LCA-Page.htm. Accessed 4 Oct 2010

33. Saouter R, Van Hoof G (2002) A database for the life-cycle assessment of Proctor and Gamble laundry detergents. Int J Life Cycle Assess 7(2):103–114

34. Blackburn R, Payne J (2004) Life cycle analysis of cotton towels: impact of domestic laundering and recommendations for extending periods between washing. Green Chem 6: G59–G61

35. Fletcher K (2008) Sustainable fashion and textiles: design journeys. Earthscan, London

36. Levi Strauss & Co (2009) a product lifecycle approach to sustainability. http://www.levistrauss.com/sites/default/files/librarydocument/2010/4/Product_Lifecyle_Assessment.pdf. Accessed 15 July 2009

37. Svoboda S (1995) Case BI: the clamshell controversy. National Pollution Prevention Center for Higher Education. http://www.umich.edu/~nppcpub/resources/compendia/CORPpdfs/CORPcaseB1.pdf. Accessed 30 July 2009

38. Good Is (2008) What Nau? http://www.good.is/post/what-nau. Accessed 5 Jan 2011

39. Wood, S (2008) Portland-based clothier with a conscience Nau in business again. The Oregonian. http://www.oregonlive.com/environment/index.ssf/2008/06/portlandbased_-clothier_with_a.html. Accessed 5 Jan 2011

40. Oppenheimer, L (2009) New owner breathes new life into Nau. The Oregonian. http://www.oregonlive.com/business/index.ssf/2009/07/new_owner_breathes_new_life_in.html. Accessed 5 Jan 2011

41. Srivastava SK (2007) Green supply-chain management: a state-of-the-art literature review. Int J Manag Rev 9(1):53–80

42. Bevilacqua M, Ciarapica FE, Giacchetta G (2008) Design for environment as a tool for the development of a sustainable supply chain. Int J Sustain Eng 1(3):188

43. Hagelaar G, van der Vorst J (2002) Environmental supply chain management: using life cycle assessment to structure supply chains. Manag Rev 4:399–412

44. Richter K, Dobos I (1999) Analysis of the EOQ repair and waste disposal problem with integer setup. Int J Prod Econ 59:463–467

45. Inderfurth K, De Kok AG, Flapper SDP (2001) Product recovery in stochastic remanufacturing systems with multiple reuse options. Eur J Oper Res 133(1):130–152

46. Chinander KR (2001) Aligning accountability and awareness for environmental performance in operations. Prod Oper Manag 10(3):276–291

47. Haas DA, Murphy FH (2003) Compensating for non-homogeneity in decision-making units in data envelopment analysis. Eur J Oper Res 144(3):530–544

48. Majumder P, Groenevelt H (2001) Competition in remanufacturing. Prod Oper Manag 10 (2):125–141

49. Kiesmüller GP, Scherer CW (2003) Computational issues in a stochastic finite horizon one product recovery inventory model. Eur J Oper Res 146(3):553–579
50. McLeod FN, Cherrett TJ (2009) Quantifying the environmental benefits of collection/delivery points. OR Insight 22:127–139
51. Fortune 500 (2010) Global 500. Our annual ranking of the world's largest corporations. http://money.cnn.com/magazines/fortune/global500/2010/index.html. Accessed 4 Jan 2011
52. Plambeck E, Denend L (2008) The greening of Walmart. Stanford Soc Innov Rev 6(2):53–59
53. Walmart (2011a) Sustainability index. Company website. http://walmartstores.com/Sustainability/9292.aspx. Accessed 4 Jan 2011
54. Walmart (2009) 2009 Walmart global sustainability report. http://walmartstores.com/sites/sustainabilityreport/2009/. Accessed 4 Jan 2011
55. Walmart (2011b) Walmart supplier sustainability assessment. http://walmartstores.com/download/4055.pdf. Accessed 5 Jan 2011
56. Report ENDS (2009) IKEA struggles to source sustainable timber. ENDS Rep 414:22

Principles of Green Manufacturing

5

Moneer Helu and David Dornfeld

> *Those are my principles, and if you don't like them ... well, I have others.*
>
> Groucho Marx

Abstract

The purpose of this chapter is to illuminate some of the basic principles of green manufacturing. That is, establish a framework of principles against which relevant examples can then be mapped to determine how green a system or solution is and find areas for potential improvements. This forms the structure of the discussion in following chapters where specific manufacturing processes or systems are investigated in more detail. We propose the following five principles: (1) a comprehensive systems approach must be used to evaluate and improve manufacturing processes from a green perspective, (2) the system should be wholly viewed across both the vertical and horizontal directions, (3) harmful inputs and outputs of the system to the environment and humans should be reduced or removed, (4) net resource use should be lowered, and (5) temporal effects on the system should always be considered. These are discussed in detail.

M. Helu (✉)
Laboratory for Manufacturing and Sustainability (LMAS),
Department of Mechanical Engineering, University of California at Berkeley,
1115 Etcheverry Hall, Berkeley, CA 94720, USA
e-mail: mhelu@berkeley.edu

D. Dornfeld
Laboratory for Manufacturing and Sustainability (LMAS), Department of Mechanical
Engineering, University of California at Berkeley, 5100A Etcheverry Hall, Mailstop 1740,
Berkeley, CA 94720-1740, USA
e-mail: Dornfeld@berkeley.edu

D.A. Dornfeld (ed.), *Green Manufacturing: Fundamentals and Applications*,
DOI 10.1007/978-1-4419-6016-0_5, © Springer Science+Business Media New York 2013

5.1   Introduction, Background, and Technology Wedges

This chapter adds structure to the discussion of specific manufacturing process or
system analysis for reduced impact. Chap. 1 provided a basic introduction to green
manufacturing and discussed its importance in the context of the motivation, basics,
and definitions associated with green manufacturing and sustainability. Much of the
discussion presented in Chap. 1 on differentiating green manufacturing from
sustainable manufacturing in terms of impact or effects was based on the concept
of "technology wedges." The paper entitled "Stabilization Wedgesdquo: Solving
the Climate Problem for the Next 50 Years with Current Technologies" published
in *Science* in 2004 describes the concept of "stabilization wedges" with respect to
technology that could solve the climate problem (with respect to fossil fuels) for the
next 50 years using current technology [1]. This idea is based on the concept of
"closing the gap" as discussed in Fig. 1.2 in Chap. 1. The "gap" discussed in the
stabilization wedges paper is the difference between the current trends in fossil fuel
emissions (so-called business as usual, BAU) and the atmosphere's capability to
accommodate emissions. This is a similar gap to the one defined earlier in Fig. 1.2
relative to sustainable design and manufacturing (normal consumption vs. sustainable consumption). Pacala and Socolow [1] propose a set of "wedges" wherein each
individual wedge represents the ability of some existing current technology to
reduce on its own some portion of fossil fuel emission. Then, summed together,
these wedges provide the necessary reduction in emissions to achieve an overall
"sustainable" situation.

This strategy can be employed in manufacturing to accomplish environmentally
benign manufacturing processes as discussed in Chap. 1. There is, as noted earlier, an
increasing interest in developing process enhancements that contribute to reducing
an environmental impact. As yet, there is no strategy to coordinate a set of enhancements and new capabilities that will, combined, render a process "sustainable" as in the
case of the fossil fuel emission example discussed earlier in Chap. 1.

There are some interesting possibilities with the development of new
manufacturing technologies for microscale and nanoscale manufacturing and
various alternate energy sources being pursued (from fuel cells to photovoltaics).
Attention will need to be paid to ensure that new processes will have a positive
effect in an environmental sense. That is, wedges must be designed to be "net-
positive." An improvement in one element of a process or system of manufacturing
cannot be at the expense of another segment of the cycle. This is especially
complicated with the complex supply chains employed today.

Continuous improvement is as valid here as in other areas of manufacturing.
Industry strives to remove "wasted time and effort." So, why not also try to do this
for energy/consumables/waste? As a background to the discussion of principles of
green manufacturing, the concept of "technology wedges" as introduced in
Chap. 1—analogous to "stabilization wedges"—offers an approach and potential
metric for addressing these energy challenges. The specifics of that metric need a
lot of development based on discussion in the community.

There are a number of fundamental rules that govern how wedge technology can be employed in manufacturing. These were explained in a paper entitled "Technology Wedges for Implementing Green Manufacturing" [2] and are as follows:

Rule 1. The cost of materials and manufacturing (in terms of energy consumption and Green House Gas (GHG) emissions, etc.) associated with the wedge cannot exceed the savings generated by the implementation of the wedge (or wedges) over the life of the process or system in which it is employed.

Rule 2. The technology must be able to be applied at the lowest level in the process chain.

Rule 3. The cost and impact of the technology must be calculable in terms of the basic metrics of the manufacturing system and the environment. That is, cost and impact must be expressible in units of dollars (or euros, yen, yuan, etc.), carbon equivalent, global warming gas creation or reduction, joules, cycle time and production rate, quality measures, lead time, working capital, and so on relative to present levels of consumption, use, time, etc.

Rule 4. The technology must take into consideration societal concerns along with business and economy, see [3] for example, and

Rule 5. There must an accompanying analytical means or design tool so that it can be evaluated at the design stage of the process or system. It must be an integrated approach.

These rules, when applied, will insure a balanced and honest appraisal of the impact of technology from an environmental perspective and also insure the rules are feasible from a business perspective. It also avoids creating anomalies in the supply chain caused by a local gain, which yields a global net loss. Many examples of a net loss currently exist in society, such as with so-called high tech/low cost products, some of which operate efficiently, but which have short useful lives.

The consumer "learns" to expect this and, thus, expects to replace the product in a short time. This trend destroys a tremendous amount of product value and resources and, worse, it encourages increased environmental damage. The life cycle of such "throw-away" products is often not considered in the design, fabrication, or use of said products

In order to effectively apply technology wedges and other approaches to greening manufacturing, a set of principles should be followed to insure positive benefit. These are laid out in the next section.

5.2   Principles

To apply the stated definition for green manufacturing and effectively use strategies such as technology wedges to real systems and solutions, a framework of principles must be established. All relevant examples can then be mapped against the defined framework to determine how green the system or solution may be as well as find any areas for potential improvements. Generally speaking, this framework takes the form of the following five principles:

1. A comprehensive systems approach must be used to evaluate and improve manufacturing processes from a green perspective.
2. The system should be wholly viewed across both the vertical and horizontal directions.
3. Harmful inputs and outputs of the system to the environment and humans should be reduced or removed.
4. Net resource use should be lowered.
5. Temporal effects on the system should always be considered.

5.2.1   First Principle: A Comprehensive Systems Approach must Be Used to Evaluate and Improve Manufacturing Processes from a Green Perspective

A comprehensive systems approach is required to consider the environmental impacts of manufacturing as these impacts could be due to any aspect of the process itself. For example, we can consider machining. As Fig. 5.1 shows, the impact of machining exceeds the electrical energy required to power the process and the cutting fluid used to cool and lubricate the workpiece. Other important aspects that are not typically considered but also do contribute to environmental impact include lubricating oil use and disposal, water use and disposal, compressed air use, and even the machines and tools themselves, which both have some amount of embedded impact due to their manufacture.

A comprehensive systems approach is also vital in the sense that most manufacturing processes are inherently complex. Continuing the machining example, while much of the work in the literature has provided a substantial base of knowledge from the process-level perspective, very little research has studied the environmental ramifications of the entire manufacturing process [4]. But, these ramifications can be significant considering the interplay at work in many manufacturing processes. For example, the trend towards MQL and dry machining provides environmental benefit by reducing the amount of cutting fluid consumed and subsequently disposed. MQL and dry machining, though, may also result in greater compressed air use or increased electricity consumption due to the increased cutting forces generated by the increased frictional interaction at the tool-chip interface or cleaning requirements to remove chips and contamination. Furthermore, other environmental aspects may be highlighted such as the increased human health toxicity due to vaporized cutting fluid in the air. Given that any one of these new factors may themselves cause substantial environmental impact, it is important for us to not only consider the effect of any one process-level improvement on the environment, but also to consider the effect of this change on other parameters that can also affect the environment.

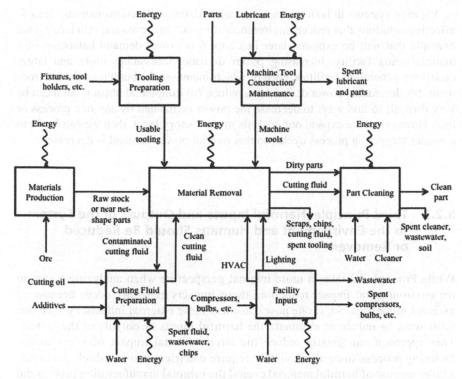

Fig. 5.1 Comprehensive systems view of a standard machining process

5.2.2 Second Principle: The System Should Be Wholly Viewed Across Both the Vertical and Horizontal Directions

The first aspect of the comprehensive systems approach advocated by Principle #1 is to view the system in both the vertical and horizontal directions where "vertical" refers to considering the system at varying levels of detail from the enterprise down to the process and "horizontal" refers to considering the system at any one level of detail. This type of approach is vital as environmental impacts can occur or be magnified depending on the level of detail one considers for analysis. An example that considers the vertical direction of a system is the energy consumption of machine tools at the factory level. While the energy consumption of a single machine tool may seem negligibly small, at the factory level this consumption may be quite large due to the number of machines in the facility. Another example that considers the horizontal direction of a system is compressed air use in a facility. Air is typically compressed using one or a small number of compressors shared by all machines and process. So, the entire machine-level of detail should be considered when analyzing compressed air usage.

Viewing systems in both the vertical and horizontal directions can also lead to effective solutions that reduce environmental impact and promote efficiency. One example that will be explored later in Chap. 6 is power demand balancing of a manufacturing facility. Increasing power demand necessitates more and larger electricity generation facilities, each of which increases the burden on the environment. So, decreasing power demand can reduce environmental impact, but it can be very difficult to find ways to decrease the power demanded by any one process or tool. However, if we expand our analysis to the factory level, then we can begin to consider staggering process cycles so that overall power demand is decreased.

5.2.3 Third Principle: Harmful Inputs and Outputs of the System to the Environment and Humans Should Be Reduced or Removed

While Principle #2 takes a more internal perspective when analyzing a system for environmental impact reduction, these impacts generally occur because of external influences. So, if one now considers these external influences, then one must work to reduce or eliminate the harmful inputs or outputs of the system. This approach can greatly reduce the environmental impact of every manufacturing process since all inevitably require electrical power, which can create a large amount of harmful material even if the original manufacturing process did not use or emit any toxic substances.

The reductions required for Principle #3 can be affected by a variety of means. If we first focus on inputs to the system, then one solution is to replace harmful inputs with other inputs that have lower impact. An alternative would be to find ways to implement recycle, reuse, or remanufacture techniques to reduce the harmful inputs required. Machining actually provides us with an excellent recycling example: cutting fluid is cleaned within the machine and then reused in order to significantly reduce the amount of cutting fluid required for the process. Similar means can be considered for harmful outputs to the system. Recycle, reuse, and remanufacturing techniques, such as the given machining example, can also serve to reduce harmful outputs from the system. Optimizing and/or redesigning the manufacturing process can also reduce harmful outputs. One good example of this approach is the use of NF_3 to clean chemical vapor deposition chambers—because NF_3 is 99% consumed, less of it is released from the system than the other perfluorinated compounds previously used. Generally speaking, though, the approach used by industry to reduce harmful outputs from a system involves the use of abatement techniques to clean any emissions as well as take-back programs to prevent any harmful outputs from unintentionally reaching any part of the environment.

5.2.4 Fourth Principle: Net Resource Use Should Be Lowered

Principle #3 can also be expanded to include all resources, even those that are not necessarily harmful but whose use can have a significant effect on the environment. For example, paper is not a harmful input to a system, but its use represents a tree that was cut down and processed releasing carbon into the environment. Our goal should be to lower net resource use. This goal should apply equally to all resources. All resources are limited and overuse of any one resource can have detrimental and potentially irreversible effects on the environment.

Ideally, net resource use should be zero so that resources may be used at a rate equal to the rate of replenishment in the environment. This statement should not be interpreted to mean that resources should be freely used since, aside from being practically impossible, it would represent inefficiency in the system that would not minimize environmental impact. There are several methods that can be employed to attempt to reduce net resource use including all of the methods given for Principle #3. One interesting example that will be discussed in Chap. 10 is the lightweighting of plastic packaging, which reduces net resource use by requiring less plastic material, less manufacturing consumables, and less electrical energy to process.

5.2.5 Fifth Principle: Temporal Effects on the System Should Always Be Considered

The previous principles were all spatially oriented—that is, they considered physical factors independent of time. However, the environmental impact of any system can vary with time. So, we must also consider temporal effects on the system. By "temporal," we are referring to any effect that is time based. These effects are inherent to the life cycle of every product or process since emissions and environmental impacts can occur at any stage of the life of a product or process. Furthermore, these emissions are typically not constant throughout the life either. A combustion engine for instance may likely have greater emissions as it ages. Even at the end of life, an item may continue to leech pollutants as it decomposes in a landfill. The emissions themselves may also have temporal effects on the environment. For example, the global warming potential of greenhouse gases is dependent on both the potency and the persistence of the greenhouse gas. Recent work in the literature, such as by Yuan and Dornfeld [5] and Zhai and Yuan [6], has tried to account for the temporal effects of many pollutants through means such as temporal discounting. Ultimately, no matter the approach taken, these lifetime effects should be accounted for and reduced when considering the impact of any manufacturing process.

Temporal effects may also refer to taking action against environmental impact during the design process. It is estimated that roughly 70% of the life cycle costs of a product or process is determined during the initial design stages. As the design progresses and decisions become set, the ability to influence eventual impacts decreases substantially. Therefore, environmental impacts should be considered as early as possible in the design stages so as to allow the greatest ability to consider alternatives that may reduce any future effects.

5.3   Mapping Five Principles to Other Methods and Solutions

Now that a detailed framework that defines the principles of green manufacturing has been established, one is in a better position to evaluate and improve methods and solutions presented in the literature. As an example, consider the five options for sustainability presented by Allwood [7]. Using the framework we have established, one can improve on Allwood's formulation as follows:

1. Use less material and energy
2. Substitute input materials (nontoxic for toxic, renewable for nonrenewable)
3. Reduce unwanted outputs (cleaner production, industrial symbiosis)
4. Convert outputs to inputs (recycling and all variants)
5. Change structures of ownership and production (product service systems, supply chain structure)
6. Apply options 1 through 5 to across all levels of the defined system
7. Apply options 1 through 5 across the entire life cycle of the system

In this example, Principle #1 is inherent in all six options that are offered since each option is based on a comprehensive systems approach. Principle #2 is mostly addressed by option 6, which dictates that solutions should be found within any level of detail of the system. While not reflecting all aspects of Principle #2, option 5 is based on the approach advocated by Principle #2 since it requires viewing the system in a different perspective. Principle #3 can be found in options 2 and 3, which both require the elimination of any harmful or toxic input or output. Principle #4 is reflected in options 1, 2, and 4 as all deal with net resource usage. Finally, Principle #5 is mainly addressed by option 7, but option 2 can also be viewed as a potential temporal issue since renewable sources are defined as such because they are not depleted in time indicating that relatively no environmental impact occurs in the future. Based on this analysis, we should be able to state that the modified Allwood formulation given would satisfactorily offer a green manufacturing system as we have defined.

5.4   Concluding Remarks

A framework that can be used to apply green manufacturing principles to existing or developing methods and solutions has been presented. It highlights the need for a comprehensive systems approach that takes into account both spatial and temporal factors in order to reduce the overall environmental impact of manufacturing across the entire life cycle of the process, equipment, and products. This framework was mapped to several existing solutions presented in the literature, and initial applications were developed that can be directly applied to achieve real, sustainable improvements.

The following chapters present several case studies that highlight these principles being applied to machining and production equipment and processes, semiconductor manufacturing, and nano-manufacturing. These case studies are followed by a discussion of enabling technologies for green manufacturing including clean energy technologies for power generation, packaging, and process monitoring and interoperability. Through these various case studies, it is hoped that the reader can better understand how to implement the principles and discussions have developed thus far to real world applications.

References

1. Pacala S, Socolow R (2004) Stabilization wedges: solving the climate problem for the next 50 years with current technologies. Science 305:968–972
2. Dornfeld D, Wright P (2007) Technology wedges for implementing green manufacturing. NAMRI Trans 35:193–200
3. Hawken P (1993) The ecology of commerce. Harper Collins, New York
4. Dahmus J, Gutowski T (2004) An environmental analysis of machining. Proc. 2004 ASME Intl. mechanical engineering congress and RD&D expo, 13–19 November 2004, Anaheim, California
5. Yuan C, Dornfeld D (2009) Embedded temporal difference in life cycle assessment: case study on VW Golf A4 car. Proc. IEEE Intl. Symp. on sustainable systems and technology, 18–20 May 2009, Phoenix, Arizona
6. Zhai Q, Yuan C (2010) Temporal discounting for life cycle assessment: perspectives and mechanisms. Proc. 17th CIRP Intl. Conf. on life cycle engineering, 19–21 May 2010, Hefei, China
7. Allwood J (2005) What is sustainable manufacturing? Sustainable manufacturing seminar series, University of Cambridge, 16 Feb

Closed-Loop Production Systems

6

Athulan Vijayaraghavan, Chris Yuan, Nancy Diaz,
Timo Fleschutz, and Moneer Helu

> *Eco-industries consume raw materials, but they try to recycle everything else and dispose as little as possible. Everything is a closed loop.*
>
> Erik Monge

Abstract

This chapter discusses the closed-loop aspects of production systems in the context of green and sustainable manufacturing. Specifically, we consider the life cycle of production systems from design and construction through use, decommissioning, and recycling or repurposing. We discuss resource and economic efficiency and present a series of examples of life cycle analysis of manufacturing systems. We also describe how to design systems for reduced

A. Vijayaraghavan
System Insights, 2560 Ninth Street, Suite 123A, Berkeley, CA 94710, USA
e-mail: athulan@systeminsights.com

C. Yuan
Department of Mechanical Engineering, University of Wisconsin-Milwaukee, Milwaukee,
WI 53201, USA
e-mail: cyuan@uwm.edu

N. Diaz • M. Helu (✉)
Laboratory for Manufacturing and Sustainability (LMAS), Department of Mechanical
Engineering, University of California, 1115 Etcheverry Hall, Berkeley, CA 94720, USA
e-mail: ndiaz@berkeley.edu; mhelu@berkeley.edu

T. Fleschutz
Department of Assembly Technology and Factory Management, Institute for Machine Tools
and Factory Management (IWF), TU Berlin PTZ2, Pascalstrasse 8-9, 10587, Berlin
e-mail: fleschutz@gmail.com

D.A. Dornfeld (ed.), *Green Manufacturing: Fundamentals and Applications*,
DOI 10.1007/978-1-4419-6016-0_6, © Springer Science+Business Media New York 2013

life cycle impact. Examples include comparisons of different machine tool systems, process parameter optimization, consumable utilization, plant services, and plant design.

6.1   Life Cycle of Production Systems

6.1.1   Background

You may wonder: who is Erik Monge? Mr. Monge is a manager of a company in Superior, Wisconsin called Elkhorn Industies that has led the development of sustainable production and products based on the forest products industry in Northern Wisconsin. It is the small companies like Mr. Monge's that are in the vanguard.

A life cycle describes the life time of an asset and is divided into several phases dependent on the perspective taken. From a market perspective, the phases are introduction, growth, maturity, and decline. A customer distinguishes purchase, operation, support, maintenance, and disposal. A manufacturer of a consumer product thinks of product conception, design, product and process development, production, and logistics [1]. For production equipment, typical life cycle phases are design, production, system configuration, operation, and disposal [2]. The life expectancy of production equipment can be determined by the following factors [3, 4]:

- Functional life: period over which the need for the equipment is anticipated
- Physical life: period over which the equipment physically exists
- Technological life: period over which technical innovation makes the equipment obsolete
- Economic life: period over which the equipment can be operated profitably
- Social and legal: period until social necessity or legal requirement dictates replacement

Facing climate change, population growth, and limited nonrenewable resources, the global economy has to adapt to a closed-loop economy [5–7]. To increase eco-efficiency, the life cycle of production systems has to be extended until the end of their physical life, and components and materials have to be remanufactured and recycled.

The increasing number of product variants, smaller lot sizes, accelerated time to market, and shorter life cycles of products have increased requirements on production equipment. Production equipment must synchronize its life cycle with the product life cycle, especially for high-volume, long production run products (Fig. 6.1). Concepts that allow a high degree of flexibility with respect to variants,

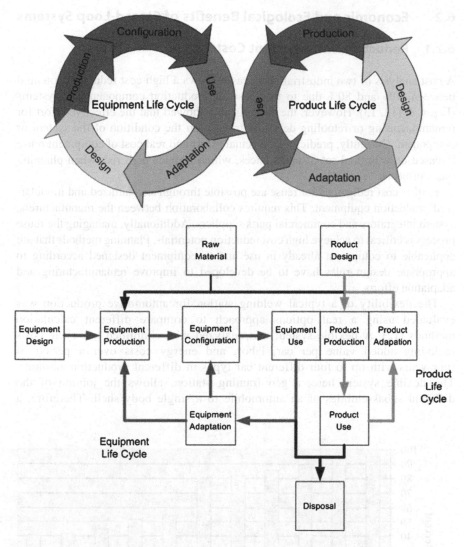

Fig. 6.1 Synchronized life cycles of products and production equipment

low-cost adaptability of products, and quick amortization within sustainable equipment use are required [8].

The current state of technological solutions and organizational practices fulfills many of the underlying requirements for the successful transfer of equipment into further use phases. Work has shown that the technical realizations of reconfigurable and transformable factories are already implemented in production facilities [9, 10]. Takata et al. [11] emphasized the potential of maintenance oriented production strategies. Methods and technologies will now be proposed that exploit the potential of production equipment.

6.2   Economic and Ecological Benefits of Closed-Loop Systems

6.2.1   Reduction of Investment Costs

A cost analysis of two industrial reuse cases shows a high cost reduction potential between 30% and 80% due to the reuse of production components or systems (Fig. 6.2) [12, 13]. However, the analysis also showed that the effective effort for remanufacturing or retooling depends strongly on the condition of the system or component. Currently, predicting the actual effort and real cost of equipment reuse is based on a limited set of experiences, which implies high risk when planning reuse projects.

Further cost reductions for reuse are possible through standardized and modularized production equipment. This requires collaboration between the manufacturers, system integrators and commercial parts suppliers. Additionally, managing the reuse process is critical to achieve high cost reduction potentials. Planning methods that are applicable to equipment already in use and to equipment designed according to appropriate design rules have to be developed to improve remanufacturing and adaptation efforts.

The flexibility of a typical welding station for automotive production was evaluated using a real options approach to compare different calculation methods [12]. A flexible system was analyzed with simplified life cycle costs including added value per car, labor, and energy costs over a period of four years with up to four different car types in different production amounts. The flexible system here, a geo framing station, allows the joining of the different subassemblies of an automobile to a single body shell. Therefore, a

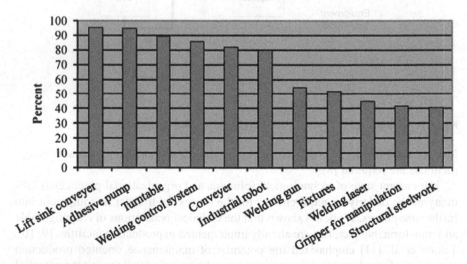

Fig. 6.2 Selected cost reductions achieved due to the reuse of assembly equipment in two industrial case studies[13]

Table 6.1 Material intensities for articulated industrial robot [14]

Articulated robot			Abiotic material		Water		Air	
Material	Unit	Amount	MI-factor	kg	MI-factor	kg	MI-factor	kg
Steel	kg	828	7.63	6,318	56	46,368	0.41	343
Stainless St.	kg	63	17.94	1,130	240.3	15,139	3.382	213
Copper	kg	250	179	44,768	236.39	59,098	1.16	290
Aluminum	kg	219	8.11	1,776	234.1	51,268	2.93	642
PVC	kg	85	3.47	316	305.3	25,951	1,703	232
Electronic	kg	20	436	8,720	5,971	119,420	264	5.28
Sum	kg	1,465		63,028		313,061		1,725
Electricity	MWh	60	1.55	277,939	66.7	4,914.201	0.54	35,482

complex frame specific to a car model with several grippers has to fix the different parts during the welding process. In contrast to a dedicated system, the flexible system can easily exchange these frames and thus allow the joining of different car models in the same stations. So, part of the investment cost for the flexible system will also apply towards new frames when new car models are introduced in the future.

6.2.2   Increase of Resource Efficiency

We may estimate the ecological reuse potentials of an articulated industrial robot as a highly flexible system and an assembly automation system as a specified productive system [14]. The evaluation was conducted using the material input per service (MIPS) unit approach developed by the Wuppertal Institute in Germany [15] and the life cycle assessment method defined in ISO 14040 and ISO 14044. Table 6.1 shows the result of the MIPS calculations.

The estimation of ecological reuse potentials is based on a standard articulated industrial robot with a possible pay load of 200 kg (e.g., the COMAU NH3). The embedded materials and their contribution to the overall weight of 1.5 tons can be seen in the first two columns of Table 6.1. The remaining columns show the material intensities per kg of the respective material regarding abiotic resource, water, and air as well as the material intensity of the material amount embedded in the evaluated robot. The second to the last line shows the sum over all materials examined. The last line shows the material intensity of the consumed energy during four years of operation time by the robot. The estimation of the energy consumption is based on measurements for standard movements assuming a power demand of 4,875 W during operation and 945 W during standby. A comparison of the material intensities shows that

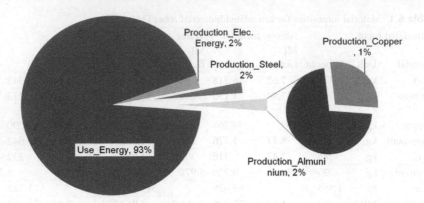

Fig. 6.3 The largest contributors to the global warming potential of an articulated industrial robot

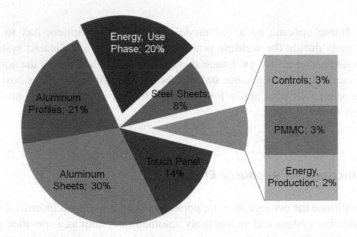

Fig. 6.4 The largest contributors to the global warming potential of an automated assembly station

the energy consumption during the use phase is the dominant and hence the principal driver of ecological impact.

Figure 6.3 shows the results of the LCA study performed using GABI software from PE International, Stuttgart. The LCA study is based on the life cycle inventory of the whole product life cycle and the results are calculated based on the estimated impacts of this inventory. In order to model a product or system the processes during the production, use, and recycling of the system are modeled with the respective resource inputs and outputs. The results are based on internationally accepted impact categories by the Centre for Environmental Studies (CML) at the University of Leiden [16]. These impact categories characterize the single in- and outputs to different environmental impacts.

Table 6.2 Material intensities for assembly automation station

Assembly automat			Abiotic material		Water		Air	
Material	Unit	Amount	MI-factor	kg	MI-factor	kg	MI-factor	kg
Aluminum	kg	338.6	19.0	6,426.6	539.2	182,573.1	5.9	2,000.8
Steel	kg	336.9	9.3	3,139.9	81.9	27,592.1	0.8	260.1
ABS, PVC	kg	7.0	5.5	38.7	205.0	1,441.2	4.2	29.5
Stainless St.	kg	5.4	17.9	96.9	240.3	1,297.6	3.4	18.3
Copper	kg	8.7	179.1	1,561.5	236.4	2,061.3	1.2	10.1
SBR	kg	0.5	5.7	2.7	146.0	69.2	1.7	0.8
Electronics	kg	5.0	174.4	872.0	2,388.4	11,942.0	105.6	528.0
Sum	kg	702		12,138		226,977		2,848
Electricity	MWh	10.0	4.7	47,000	83.1	831,000	0.6	6,000

Figure 6.4 and Table 6.2 show the respective results for the automated assembly station with half a shift use per day over four years. The evaluated assembly station is a packaging machine consisting of two conveyer belts, four pneumatic pick and place handling devices and two pneumatic centering devices. The calculations are based on the bill of material and energy based on compressed air field measurements. In contrast to the robot, the energy consumption during the use phase only represents 20% of the overall global warming potential in the LCA study. Thus the composition of the station and the embodied materials, mainly aluminum and steel, influence the life cycle ecological impact.

The material intensity (MIPS) and LCA results were substantially different in these two case studies. Some of these differences can be explained by the fact that the MIPS approach does not consider the end of life phase and the data for energy consumption seems to be quite different. Unfortunately, the calculation method and assumptions behind the MIPS data is not public, and thus a further analysis of these difference is not possible. Nevertheless, the variations seen in this case study seem to indicate that the difference is too big to use MIPS as proxy for the ecological impact. The evaluation of the robot shows that the energy consumption during the use phase has to be investigated in more detail to understand the ecological impact more precisely. The following section will present a study of machine tools to predict the energy consumption of production equipment.

6.3   Machine Tools and Energy Consumption

The focus in this section is closer to the production floor at the line/cell level with specific interest in one of the primary pieces of manufacturing hardware: the machine tool.

Machine tool energy consumption may be reduced in one of four areas of its life cycle: manufacturing, transportation, use, or end of life. Early life cycle assessments of machine tools and manufacturing processes have focused on quantifying the energy and resource consumption during use. Shimoda [17] contended that minimizing cutting fluid consumption provides a more effective means of saving energy when compared to the use of recycled material in manufacturing a machine tool since the use of the machine tool is extensive. However, Diaz et al. [18] showed that the impact of the manufacturing and transportation of the machine tool with respect to carbon-equivalent emissions per part produced depended on the facility in which the machine tool was used. Much of the literature on machine tools and the environment reduces the scope of the analysis and presents design- or process-level changes, each of which affects the energy requirements of the machine tool during its manufacture and use.

Design-level changes provide the greatest flexibility and therefore the greatest opportunity for energy savings [19]. Such strategies include design for disassembly [20, 21] and remanufacturing to reuse material for the machine tool frame [22]. Strategies that require a design change of the machine tool to save energy during use are also extensively studied, such as Minimum Quantity Lubrication (MQL), which provides the added benefit of using three to four times less cutting fluid than conventional flood cooling [23]. MQL strategies require modifications to the cooling system of the machine tool if it uses an internal coolant feed, though [24]. Dry machining has been another area investigated to eliminate the impacts of cutting fluid. While dry machining does not require substantial machine tool design changes, proper tooling and cutting conditions must be practiced to reduce excess tool wear, which would overshadow initial energy savings [25].

Munoz and Shengy [26] developed a model that incorporated cutting fluid flow as an environmentally conscious measure in machining as well as process-level dynamics such as machining mechanics and tool wear. This model served as the foundation for the development of an environmental process planning system that works with conventional process planning methodologies to evaluate trade-offs between environmental and productivity requirements [27]. Narita et al. [28, 29] developed a similar tool called an "environmental burden calculator" related to part manufacture that allowed a user to input cutting conditions and work piece information.

Recent research also includes power consumption analyses of machine tool use. Dahmus and Gutowski [30] conducted an environmental analysis of machining that quantified the energy consumption of four types of milling machines varying in automation as well as accounted for material production and cutting fluid preparation. Taniguchi et al. [31] studied the effects of downsizing a CNC milling machine tool on its energy and resource requirements. Gutowski et al. [32] broadened the scope to include ten types of manufacturing processes, and noted that the low throughput of additive processes such as sputtering amplify the specific energies relative to other manufacturing processes even though the power requirements of the processes studied do not vary by more than 2 orders of magnitude.

Ultimately, a life cycle energy assessment is required to determine the appropriate strategy to "green" a machining process. This type of analysis yields two general possibilities: (1) high constant energy demand due to the dominance of noncutting operations and peripheral equipment, or (2) low constant energy demand due to the dominance of cutting operations. Sample strategies to address the first case include using machine design to minimize the energy requirements of peripheral equipment (e.g., kinetic energy recovery systems used in conjunction with the spindle) and focusing on machine operation to increase the production rate of the machine tool. Strategies to address the second case generally require optimization of the cutting process itself. This may be difficult to accomplish from a design perspective due to the influence of desired process parameters, but energy savings may be achieved by considering typical machine tool use in design (e.g., ensure that axes with high motion carry less weight).

6.4   LCA of Machine Tools

While the current literature provides an extensive knowledge of the life cycle energy consumption of machining, it is limited by the assumption that machine tool operation dominates the overall impact such that other aspects of the machine tool's life cycle, such as its manufacture, are neglected. Furthermore, much of the literature neglects transportation, material inputs (e.g. cutting fluid), or facility inputs (e.g. HVAC and lighting), which may all have a significant impact on the overall energy consumption. So, it was the goal of Diaz et al. [18] to study the effect of these aspects as well as that of the manufacturing environment and degree of automation on the life cycle energy requirements of milling machine tools. The work is summarized here as an example of a systematic analysis of process technology and equipment.

6.4.1   Methods

Two types of machine tools were studied in this analysis: (1) the Bridgeport Manual Mill Series I (for low automation), and (2) the Mori Seiki DuraVertical 5060 (for high automation). Energy consumption and CO_2 emissions were calculated for each life cycle stage in different manufacturing environments [18].

Each machine tool was divided into its primary components (machine tool frame, spindle, ball/lead screws, X/Y axes, tool changer, casing, and controller) to determine the energy consumed during its manufacture. The material composition of the components were simplified—the machine tool frame was assumed to be gray cast iron, the casing was low carbon steel, and the remaining components were low alloy steel—and all choices assumed standard recycling content [33].

The following processes were considered when calculating the energy consumed during the production of each component: casting, extrusion, rolling, stamping, milling, turning, grinding, case hardening, annealing, and tempering. Embodied energy of deformation processing was used for the extrusion, rolling, and stamping processes [33]. Specific energies were used for the milling, turning, grinding, case hardening, annealing, and tempering processes [32, 34–36]. To compute resultant

CO_2 emissions, a Japanese energy mix (360 g of CO_2-e/kWh) was used for the Mori Seiki [33] and a Connecticut energy mix (420 g of CO_2-e/kWh) was used for the Bridgeport [37, 38].

Transportation energy and CO_2 emissions were calculated—the Mori Seiki originated in Nagoya, Japan, and the Bridgeport originated in Bridgeport, CT [33]. Both were sent to San Jose, CA for use and then to Los Angeles, CA for resale at the end of life.

To analyze the effect of different facility characteristics and production schedules, the use phase of both machine tools was studied across three manufacturing environments: a community shop, a job shop, and a large commercial facility. The functional unit of a machine tool in each environment depends on its performance and ends once resold by the original owner. A 101 × 101 × 25.4 mm AISI 1018 steel standard part served as the functional unit in this analysis.

Energy consumption was measured during part production. Cutting fluid was considered for both machine tools, while lubricating oil was only considered for the Mori Seiki; both analyses utilized an embodied energy approach. The energy required for HVAC and lighting to support machine tool operation was calculated based on facility square footage and data from [17]. Total HVAC and lighting energy was allocated to the machine tools according to the size of the workspace required to operate the tool. Emissions were calculated using a California energy mix (320 g of CO_2-e/kWh) [38–42].

Labor and workpiece preprocessing were omitted. An end-of-life analysis has also been omitted due to the uncertainty in the amount of times a machine tool is reused. But, material recyclability was accounted for when considering the manufacture of the machine tool.

6.4.2 Results

The energy required to manufacture the Bridgeport and Mori Seiki was estimated to be 18,000 MJ and 100,000 MJ per machine tool, respectively. Material extraction was the most energy intensive process—it was responsible for 70% of the total energy consumed in manufacturing for both machine tools—followed by casting. Accordingly, the machine tool frame was the component of both machine tools that required the greatest amount of energy to manufacture.

Both machine tools have similar transportation emissions; 1,200 kg of CO_2-equivalent for the Bridgeport and 1,600 kg of CO_2-equivalent for the Mori Seiki. Now considering the actual use of the machine tools, the Bridgeport consumed 600 kJ per part and the Mori Seiki consumed 1,000 kJ per part to manufacture the standard part that served as the functional unit. Maintenance energy consumption was negligible while HVAC and lighting consumed 40–65% of the total energy required during use of the machine tools. The most energy intensive scenario during use of the machine tool was the Mori Seiki in the community shop due to the low production volume; the energy consumed in this scenario was 2,800 kJ per part.

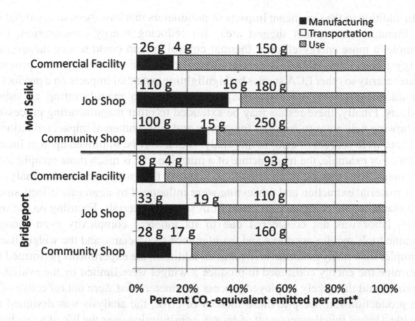

Fig. 6.5 CO$_2$-equivalent emitted per standard part produced. *Numbers provided are in grams of CO$_2$-equivalent emitted per part

The CO$_2$-equivalent emissions calculated for both machine tools in all three manufacturing environments resulted in measureable differences with the manufacture of the machine tools being significant relative to their use (see Fig. 6.5). The percentage of CO$_2$-equivalent emissions during the manufacture of the machine tools was smallest for both machine tools in the commercial facility because of the higher production rates possible. The use of the machine tools dominated the total emissions, varying from 70 to 90% of the Bridgeport's emissions and 60–85% of the Mori Seiki's emissions.

6.4.3 Conclusions

This analysis of two milling machines placed in three environments has quantified the CO2-equivalent emissions associated with producing a standardized part over the lifetime of a machine tool. Several findings show significant differences from previously published literature, such as the manufacturing phase impact. Since HVAC and lighting requirements are a significant portion of the use phase emissions, future studies should use specific facility HVAC and lighting data since average national energy intensity data were used in this study.

In addition to the significant impacts of parameters that have been disregarded in the literature, the results suggest areas for reducing energy consumption. For example, a more energy-efficient thermal control system could reduce the overall energy usage since HVAC is energy intensive. This analysis may also provide greater clarity to other LCA studies by highlighting potential impacts on a product's manufacturing phase since machine tools are key to manufacturing all other products. Finally, these results may be extended to other manufacturing processes by showing new areas to consider for energy and environmental impact reductions.

There were sources of error in this analysis that may be improved upon in future studies. For example, the manufacture of a machine tool is much more complicated with many steps that are more difficult to quantify than we present in this analysis. Also, material extraction and processing were influenced by aggregate effects since each machine tool component is made from several materials. Focusing on the use phase, labor was not considered due to its inherent complexity even though machine tools require operators and maintenance technicians, and the widget itself is simpler than many manufactured parts. In addition, the experiments performed to determine the energy consumed to produce a widget were limited by the available resources and thus likely employed process parameters that were not reflective of a true production run. Despite these sources of error, the analysis was designed to provide a broad initial assessment of energy consumption over the life of a machine tool. Future work will strive to improve the data used in this analysis as well as refine the analysis itself by incorporating other potentially significant factors.

6.5 Process Parameter Optimization

Given the significant environmental impact of the use phase of a machine tool, an analysis was conducted on end milling process parameter selection for energy consumption reduction [43]. The energy per unit manufactured is determined by both the power demand of the machine tool during machining and the processing time (see Fig. 6.6).

The power demand of a machine tool may be divided into a constant and a variable component [30]. The constant power can be attributed to the computer, fans, lighting, etc. of the machine tool. This component of the total power demand is independent of process parameter selection. The variable power demand, though,

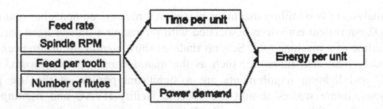

Fig. 6.6 Influence of process parameters on the energy per unit manufactured

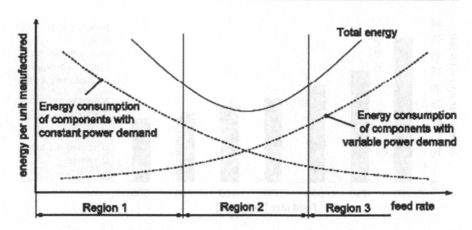

Fig. 6.7 Regions of machining process

is dependent on process parameter selection and can be attributed to the spindle or the drives of the table axes.

The processing time per unit manufactured is determined by the feed rate. Cutting conditions at a specific feed rate depend on the selection of the number of revolutions per minute of the spindle, the feed per tooth and the number of flutes.

Figure 6.7 identifies two opposing effects on the energy per unit manufactured. First, as the feed rate increases the processing time is reduced. Therefore, the contribution of the constant power demand of the machine tool to the energy per unit manufactured decreases. Second, an increase in the feed rate demands more power from the machine tool with or without adjustment of the cutting speed. Depending on which effect prevails, three different machining regions may be found.

The sum of both energy contributions results in a parabolic-total energy plot also shown in Fig. 6.7. In Region 1 the decrease due to a shorter processing time dominates the increased variable power demand. In this region the feed rate will be chosen as fast as technically possible. In Region 2 the energy per unit manufactured is fairly constant, whereas the increase of the variable power demand dominates in Region 3. If the process is located in Region 3, slower feed rates would lead to lower energy per unit manufactured.

Experimental studies were conducted to determine the region of the machining process in Fig. 6.7. Initial experiments (a) kept the feed per tooth constant by increasing the spindle speed proportionally to the feed rate and (b) varied the feed per tooth at a constant spindle speed. Subsequently, the energy per unit manufactured of conventional cutting with a 2-flute uncoated carbide end-mill was compared to the energy per unit manufactured of high speed cutting with a 2- and 4-flute TiN-coated end-mill.

Slot cutting experiments of a low carbon steel (AISI 1018) were conducted on a Mori Seiki NV1500DCG with an 8 mm uncoated carbide end-mill and a depth of

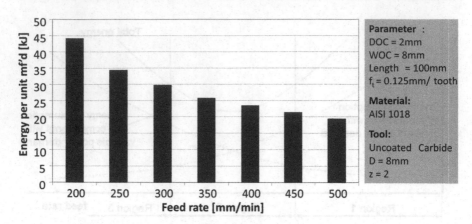

Fig. 6.8 Average energy per unit manufactured versus feed rate

cut of 2 mm to be consistent with [31]. For each process parameter combination, 52 slot cuts were performed in order to study the wear of the tool. The power demand of the machine tool was recorded using a WattNode Modbus power meter via an MTConnect monitoring system (see Chapter 11).

6.5.1 Constant Feed per Tooth

The initial cutting conditions used a recommended feed per tooth of 0.125 mm/ tooth and a spindle speed of 800 rpm to generate a feed rate of 200 mm/min. Figure 6.8 shows that the energy consumed by the machine tool per unit manufactured decreases over feed rate at a constant feed per tooth of 0.125 mm/ tooth. However, tool wear increases significantly over feed rate. The tool consistently broke before having cut 52 slots at a feed rate of 500 mm/min.

It has to be taken into account, that tool production consumes energy as well. Thus, increasing the cutting speed at a constant feed per tooth might result in smaller amounts of energy consumed by the machine tool per unit manufactured, the increased wear of the tool leads to greater overall energy demands of the process. It can be concluded that increasing the spindle speed at a constant feed per tooth is not a strategy to lower the energy demand of the material removal process.

6.5.2 Constant Spindle Speed

The feed per tooth was varied between 0.025 mm/tooth (feed rate of 40 mm/min) and 0.15 mm/tooth (feed rate of 240 mm/min). Preliminary studies showed that increasing the feed per tooth beyond 0.15 mm/tooth results in tool breakage.

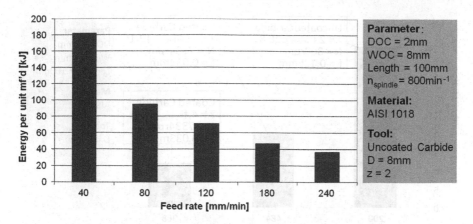

Fig. 6.9 Average energy per unit manufactured versus feed rate at constant spindle speed

Table 6.3 Summary of conventional versus high-speed cutting parameters

	Units	Conv.	High-speed
Coating	–	–	TiN
z	–	2	4
$n_{spindle}$	rev/min	800	7,334
f_t	mm/tooth	0.125	0.033
v_f	mm/min	200	968
t_{cut}	s	28.8	6.0

Figure 6.9 shows that by lowering the feed per tooth (analogous to lowering the feed rate) the energy per unit manufactured increases. However, surface quality improves creating a trade-off between surface quality and energy consumption during machining. Furthermore, the tool wear effectively decreases as the feed rate is reduced since the load on the tool is smaller.

6.5.3   Conventional Versus High-Speed Machining

High-speed cutting with coated end-mills involves greater cutting speeds at a lower feed per tooth. The feed rate increases since the increase in cutting speed is greater than the decrease in the feed per tooth compared to cutting at conventional speeds. In this study a 4-flute TiN coated end-mill was used and the energy consumed was compared to the 2-flute uncoated end-mill. Table 6.3 summarizes how the number of flutes (z), spindle speed ($n_{spindle}$), feed per tooth (f_t), and feed rate (v_f) were varied across the three experiments.

The energy per unit manufactured decreases due to the selection of a high-speed cutting process over a conventional cutting process (Fig. 6.10). The increase in power demand due to the higher spindle speed and the greater feed is compensated

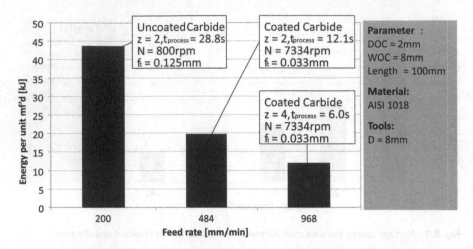

Fig. 6.10 Energy comparison of conventional cutting versus high-speed cutting

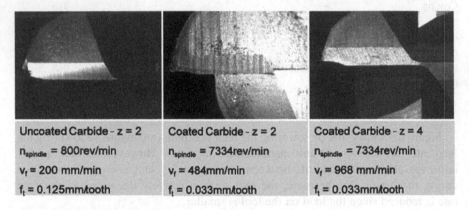

Fig. 6.11 Tool wear comparison after 52 slots

by the shorter cutting time. The selection of a 4-flute end-mill over a 2-flute end-mill reduces the energy per unit even further. The wear is reduced by choosing coated end-mills and running high-speed cutting processes.

Figure 6.11 shows the tool wear of the conventional and the high-speed cutting tools after 52 slots. It can be seen that although high-speed cutting using a TiN-coated tool leads to a greater productivity at a higher feed rate, the tool wear is reduced in comparison to the conventional cut.

If the additional production energy of providing coating to a tool is reasonably small compared to the total production energy of the tool, the fraction of the production energy of the tool that has to be considered within the equation of the energy per unit manufactured is roughly constant for conventional and high-speed cutting operations. Under this assumption high-speed cutting operations are more energy efficient than conventional cutting processes.

6.5.4 Conclusions

The tool limits the possibilities to reduce the energy per unit manufactured by process parameter selection, because elevated tool wear was observed at process parameter combinations, which reduce the energy demand of the machine tool during the manufacturing process. In order to examine the influence of tool choice, the energy per unit manufactured of TiN-coated end-mills at their recommended cutting parameters were compared to the energy per unit of uncoated end-mills. Coated end-mills enable high-speed cutting, whereas the uncoated end-mills can be used for conventional cutting only.

High-speed machining is carried out at drastically elevated cutting speeds and reduced feeds per tooth compared to conventional cutting. In general, higher cutting speeds increase the power demand of the machine tool whereas a lower feed per tooth reduces the power demand. The experimental studies show that high-speed cutting results in smaller energies per unit manufactured compared to a machining operation at conventional cutting speeds. The decrease in the processing time has a greater impact on the energy demand per unit manufactured than the increase in power demand. Also, only minimal tool wear was observed on TiN-coated end-mills after the predefined number of 52 slot cuts. Therefore, it can be concluded that high-speed cutting is more energy efficient than cutting at conventional speeds. Future work should include the variation of coating as well as the consideration of the additional production energy of coating.

6.6 Dry Machining and Minimum Quantity Lubrication

From an economical and environmental point of view, it is desirable to minimize the coolant use in machining. In the metal cutting industry, coolants play a crucial role in the production of high quality products and are used in large quantities. Old coolant or chips with oil contamination have to be specially disposed of in order to prevent contamination of the ground water. The bulk of coolants are water mixed with a concentration between 3 and 10%, such that the avoidance of coolants helps reduce the contamination of water. It is estimated that the cost of coolant lubricants is between 7 and 17% of the total manufacturing costs, depending on the process and the production location [24]. So, the minimization of coolant use will help reduce this cost component drastically. Moreover, coolants can affect the body (air way or skin disease) and elicit allergies.

Fig. 6.12 External and internal feed systems for MQL

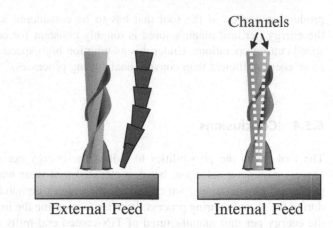

External Feed   Internal Feed

For many machining processes the use of coolant is not imperatively necessary. However, simply turning off the coolant supply can deteriorate the cutting result, since the functions of the coolant are not fulfilled any more. In detail, the primary functions of coolant lubricants comprise lubrication, cooling and cleaning. Without these functions, more friction and adhesion between the tool and the workpiece occur. Moreover, the heat produced in the cutting process has to be discharged only by the chips, the tool and the workpiece and not by the lubricant any more. The result is a higher thermal load on the tool, the workpiece and the machine tool, causing a shorter tool life and a reduced workpiece and machine tool accuracy. Therefore, the absence or minimization of coolant lubricant as with minimum quantity lubrication (MQL) necessitates an analysis of the boundary conditions and the work dependencies between the process, the cutting tool, the workpiece and the machine tool.

MQL is an alternative to the large consumption of cutting fluid during flood cooling and high temperature machining that dry cutting creates since it uses small amounts of lubrication during machining. MQL techniques have flow rates which range from 50 to 500 mL/h, which is three to four times lower than flood cooling [23]. This section will analyze how the MQL technique compares to flood cooling and dry cutting with respect to the quality of the part, tool wear, health and environmental effects, and the challenges that the user faces in implementing MQL.

MQL can be implemented by two different methods: using external or internal feeds [23]. The external feed is similar to how flood cooling is accomplished where a nozzle is adjusted so that it points at the cutting tool and workpiece interface (see Fig. 6.12). In the case of internal feed, a channel(s) is built in the cutting tool so that the coolant ejects precisely onto the cutting tool and workpiece interface, making this method more effective at cooling the tool/workpiece interface.

Experiments showed that the use of MQL while turning AISI 4340 steel resulted in a 5–10% decrease in cutting temperature when compared to flood cooling [23].

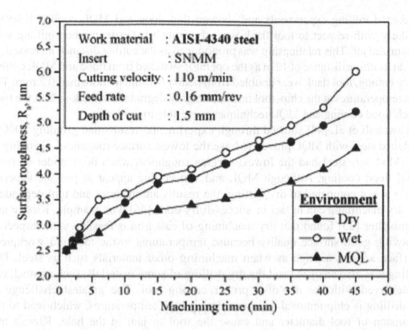

Fig. 6.13 Surface roughness for various cooling conditions [23]. Copyright 2006 Elsevier reprinted with permission

The surface roughness of the turned part also proved to be superior to dry cutting and flood cooling; the difference in surface roughness amongst the three techniques was even more disparate as machining time progressed as can be seen in Fig. 6.13 in the absence of coolant or lubrication (i.e., under dry cutting conditions).

A surface roughness analysis of the three cooling conditions was also conducted on grinding of 100Cr6 hardened steel and 42CrMo4 soft steel by [44]. In grinding, surface quality is of even greater importance since it is a finishing process. The results were consistent with [23] in grinding 100Cr6 hardened steel in that MQL proved to have a lower surface roughness than when grinding dry or with flood cooling, which was attributed to plastic deformation in the contact zone. Dry grinding of 100Cr6 hardened steel resulted in the largest surface roughness, while the surface roughness of wet grinding was in between dry and MQL. For the grinding of 42CrMo4 soft steel wet grinding fared better surface quality than MQL and dry grinding conditions.

Aside from the surface quality of the machined tool, the length of the tool life and thus the wear that the cutting tool undergoes is another important point to take into consideration when analyzing the effects of cooling conditions while cutting. Tool wear experiments comparing dry, wet, and MQL cutting have been conducted by Dhar et al. [23] and Weinert et al. [24]. Dhar et al. [23] found that using MQL while turning resulted in less wear than when flood cooling or cutting dry. Wakabayashi et al. [45]

conducted milling experiments and showed that flood and MQL cooling behaved similarly with respect to tool flank wear, but much better than when milling with pressurized air. This relationship was pronounced as the cutting distance increased. In fact, at a cutting distance of 12 m as the operator switched from flood and MQL cutting to dry cutting, tool flank wear doubled from nearly 0.1 mm to more than 0.2 mm. The high temperatures at the chip–tool interface highly degraded the flank-side of the tool, which flood cooling and MQL techniques were able to minimize.

Tawakoli et al. [44] showed through experimental results that grinding 100Cr6 hardened steel with MQL proved to have the lowest surface roughness. Grinding of 42CrMo4 soft steel had the lowest surface roughness when done under conventional flood cooling. Although MQL and wet cutting appear to provide superior results when compared to dry cutting, the results are material and tool dependent and dry machining can in fact be successfully done [46]. For example, Klocke and Eisenblatter [25] found that dry machining of cast iron is feasible with respect to achieving good surface quality because temperatures in the tool and workpiece interface are not as high as when machining other materials such as steel. Dry drilling was also analyzed and the dry drilling of some materials, such as steel, can be achieved with the use of a proper cutting tool. The greatest challenge in dry drilling is chip removal and overcoming high temperatures, which lead to the expansion of tool diameter and cause the tool to jam in the hole. Klocke and Eisenblatter [25] also illustrated that dry drilling of steel can be successfully accomplished with a TiN-coated tool with a modified flute and tapering. It was also shown that the flank wear when dry face milling the cast aluminum alloy, AlSi10Mg-wa, can be reduced if the cutting tool type was changed from an uncoated tool to a TiN+MoS2 or diamond-coated tool.

6.6.1 Health and Environmental Hazards

The surface quality and tool wear aspects have been analyzed for the various cooling conditions, but it is also important to take into consideration the health and environmental hazards. The following health and environmental risks with relation to coolant use have been identified:

Improper disposal: Since coolant is usually considered a toxic material and therefore hazardous chemical, it must be disposed of through a proper environmental, health, and safety organization. The disposal of toxic chemicals therefore becomes expensive, which is why it would be ideal for a manufacturing facility to recycle its coolant internally. If the coolant were to be disposed of improperly to the environment, the coolant could leach into the soil and/or water supply and potentially pose a significant risk to the surrounding community.

Bacteria build-up: Within a machine tool, coolant is typically stored in an enclosed tank, in the dark, and at warm temperatures due to machining, conditions ideal for the build-up of bacteria. If the coolant is not properly treated with anti-fungicide, the operator is at risk for skin disease if they have an open wound and have direct contact with the bacteria-infested coolant [47]. Bacteria would be less of a problem with MQL systems than with flood cooling because a significant amount of the coolant evaporates upon contact with the tool/workpiece interface due to the high temperatures.

Inhalation of coolant mist or particulate matter: When flood cooling is not used, the inhalation of coolant mist with MQL or particular matter with dry machining becomes more likely. The inhalation of coolant mist could be avoided by using a system as has been proposed by [48], the direct oil drop system (DOS). Particulate matter in dry grinding, for example, could be avoided by installing a ventilation system above the workspace.

The aforementioned health and environmental hazards can be avoided by following proper safety protocol, but are noteworthy nonetheless.

6.6.2 Challenges of Implementing MQL and Dry Cutting

There are of course some challenges with implementing MQL and/or dry cutting, which include the cleanability of the workpiece and the cost associated with incorporating an MQL cooling system. In conventional machine tools with flood cooling systems the coolant serves as the primary chip removal mechanism in that it carries the chips to the chip collector. With an MQL system or dry cutting, though, there would not be sufficient coolant to remove the chips. The operator would have to resort to manual removal, which means increased costs associated with additional postprocessing tasks or the incorporation of a pressurized system to automate the chip removal process. The latter situation, though, would result in the consumption of additional energy.

The second challenge associated with the MQL system requires the modification of the machine tool to install appropriate pumps and sensors. So, costs for the modification would arise, but since they could be amortized over the lifetime of the machine tool, MQL would still be a less-expensive option considering that coolant consumption would decrease tremendously. Also, the energy consumed while running MQL in comparison to flood cooling systems could be greater because of the additional sensors and special tooling (if using internal coolant feed) that would be required. But, once again less coolant would be consumed so the energy to manufacture and transport the coolant would decrease accordingly.

The additional costs of changing the cutting tool and machine tool, and the environmental and safety effects associated with MQL and dry cutting should be considered on a case by case basis when choosing which cooling method to practice for a particular machining application. Even though under conventional cutting conditions MQL has proven to be superior to dry and wet cutting in most instances, dry cutting can be successfully accomplished by changing the cutting tool accordingly to achieve the desired results.

6.7   Remanufacturing

6.7.1   Product Recovery Management

"The management of used and discarded products, components or materials within a manufacturing environment" is referred to as product recovery management [49]. The aim of product recovery management is to recover as much economical and ecological value of an asset (so-called cores) as possible and avoid as much waste as possible. Six different recovery strategies exist:

Reuse is the simplest form of product recovery. It can be defined as "...the process of collecting used materials, products, or components from the field, and distributing or selling them as used. Thus, although the ultimate value of the product is also reduced from its original value, no additional processing is required" [50]. The German Industry Standard VDI 2243 [51] differentiates the terms "reuse" and "continued use." "Reuse" is the use of an object for the same purpose, and "continued use" is the use of an object for a different purpose. Besides the reuse of a complete product, the reuse of components is possible which leads to the definition of reuse as the "process of using a functional component from a retired assembly" [52].

Repair is probably the best know product recovery process. Repair is the "process of bringing damaged components back to a functional condition", including disassembly, reassembly, and replacement of components [52]. Specific to repair (as a subprocess of maintenance) is its application not only at the end of use or end of life, but also during the complete life cycle.

Reconditioning and *refurbishment* are simultaneous terms for the "process of restoring components to a functional and/or satisfactory state but not above original specification using such methods as resurfacing, repainting, sleeving, etc." [53]. Products are disassembled, all critical components are inspected, if necessary repaired or replaced and reassembled to the reconditioned or refurbished product.

Remanufacturing can be defined as "the process of bringing a product to like-new condition through reusing, reconditioning, and replacing component parts" [53]. Lund et al. [54] also use the terms restoring of a discarded or traded-in cores to a like-new condition. Nasr and Thurston [55] add the term upward-remanufacturing if products are remanufactured for integration into new products or systems with extended functionalities. Remanufacturing encompasses collection of cores, cleaning, inspection, separation, disassembly, reprocessing, reassembly, and redistribution.

Cannibalization can be defined as "the process in which a limited number of components are extracted from a product for recovery" [56]. This differs from remanufacturing since remanufacturing preserves a product's identity and replaces or reprocesses the components while cannibalization reprocesses some components as spare part for other products.

Recycling aims to reuse materials of used products and components for the fabrication of new component parts [49]. Products for recycling are dismantled and carefully sorted and processed in order to recover pure material of high quality. The major drawback of recycling is its destruction of assed value (e.g., added labor or energy) [54].

Product recovery is typically split into six sub processes:

* *Collection*: Two different scenarios can be distinguished: end-of-use and end-of-life returns. End-of-use returns are returns of used working products for which the end-user has no further use. End-of-life returns are at the end of the products life cycle.
* *Inspection and Separation*: The cores are inspected, cleaned and separated after the collection. The aim of this process is the evaluation of the core's condition and the determination of its further treatment (e.g., disassembly level). The material flow is divided into a flow for direct reuse for product reprocessing or for final disposal.
* *Disassembly*: The recoverable cores are disassembled with the aim to selectively extract valuable and pollutant-containing components. The separated components are cleaned, inspected, tested, and evaluated for further treatment.
* *Reprocessing*: Possible operations are cleaning, replacement, machining, painting, inspection, and testing [58]. A used part or component is transformed into a reusable part or component.
* *Reassembly*: Reassembly is the process which reintegrates reprocessed parts and components into the forward supply chain. After the final functional test the product is prepared for shipment or put into storage [58].
* *Redistribution*: Redistribution is concerned with the market supply of recovered products. It includes logistics and sales operations to move recovered products to retailers or prospective end-users [59].

The literature on remanufacturing classifies remanufacturing companies as OEMs, contracted remanufacturers, and independent remanufacturers [54, 57]. Independent remanufacturers buy cores from end-users or core brokers for remanufacturing and resale. Contracted remanufacturers remanufacture products on a third company's behalf; often an OEM. They have the advantage of technical knowhow and replacement parts from the OEM [57]. Often OEMs remanufacture their own products in order to offer their customer additionally low-cost solutions. OEMs can use their service network and contact with customers to facilitate the acquisitions of cores [54]. These three are surrounded by various supply and service firms: core brokers, warehouse distributors, parts and component suppliers, process equipment suppliers, consulting services, academic research, and trade associations [54].

6.7.2   Industrial Practice in Remanufacturing of Industrial Robots

The Xerox Corporation is a well-known example of a company that uses consumer product remanufacturing as part of its business strategy. Xerox reclaims its copier at the end of its service life and returns them to market after a remanufacturing process [53]. In his study of remanufacturing companies in the US, Lund [54] could identify 48 of the 401 listed companies in the association for manufacturing technology that are offering retrofitted products. The companies belong to different fields such as metal cutting, forming machinery, mining, oilfield, material handling, textile, woodworking, food processing, and special industry machinery. Another example is Caterpillar, the world's largest maker of construction and mining equipment, diesel and natural gas engines, and industrial gas turbines, who offers certified used equipment, rebuilt power trains, and remanufactured products with like new warranties [60]. In 2007, Caterpillar remanufactured 2.2 million cores [61].

The remanufacturing of industrial robots is offered by OEM manufacturers such as Kuka Robotics, ABB Automation, or by smaller third party companies. The International Federation of Robotics (IFR) indicates that refurbished robots are gaining importance and estimated a market share of 10% in 2007. ABB is offering an "ABB Certified Refurbished Robot," with 152 check points, 12 months warranty, and ABB OEM factory specifications according ISO 001 standards and a 24 h testing period with maximum load [62]. The cost for remanufactured robots are between one-third or one-half of the replacement costs [63]. Specific to the remanufacturing of robots is the mainly complete disassembly of the manipulator and the replacement bearings and internal wiring harness if it requires harmonic drives and gears [62, 63].

6.7.3   Challenges and Opportunities in Remanufacturing

Lund [54] points out that the main driver for remanufacturing is profitability. From industrial examples, a price reduction up to 50% for remanufactured products compared with new products is possible. Additionally, the remanufacturing of a company's own products allows companies to apply new business strategies (e.g., product–service system or operator models) and protect their brand by controlling the aftermarket [56].The customer feedback gained can be used to tighten the customer relationship and can be valuable feedback for product development where the user acts as the innovator. Looking at remanufacturing from a societal view, the mainly manual remanufacturing processes encourages a higher employment rate. Finally, one cannot forget the ecological impact of remanufacturing: recovered added value and reduced energy in the initial manufacturing processes of remanufactured goods.

The main challenges of remanufacturing are small lot sizes, unknown conditions of the cores, and a poor availability of cores [64]. Further obstacles include the sourcing of appropriate replacement parts and the acquisition of technical information and documentation of the products [53]. Lund [54] adds to this the lack of public image and difficult dissemination of technical knowhow. The few R&D efforts in the remanufacturing field is likely a result of the company structure with mainly small and medium sized enterprises (SMEs) not able to finance a significant R&D program.

Nasr and Thurston [55] formulate design principles for remanufacturing. A product should be designed for disassembly, multiple life cycles, and on a modular structure. Especially fasteners and interfaces have to be adapted to facilitate the access and replacement of components. Embedded condition monitoring can support decision making in product recovery management.

6.8 Reuse

Reuse of equipment, where feasible (and distinct from remanufacturing) is another recovery strategy. Equipment used in assembly offers a good example. Assembly equipment is often used for a shorter period than its physical and technological life cycle. Industrial companies could increase their cost-efficiency and take advantage of ecological benefits by reusing their assembly equipment and thus exploiting their whole potential. The mean physical life cycle of manufacturing equipment is around 20 years [65]. In a study by Schmaelzle [66], half of the examined products were produced for less than six years. Only 11% of the examined products from assembly operations were produced for over 12 years.

Within the European Integrated Project PISA the potential of reuse was identified in a survey showing that manufacturers recognize the advantages of equipment reuse and that reuse is increasingly undertaken [13]. Nevertheless, the analysis of reasons for reuse shows that decisions about the reuse process are not specified in business decision-making processes and are not based on relevant knowledge or information. This results in costly and inefficient reuse projects. Almost every reuse case was carried out without using acquired knowledge and experience from prior and successful reuse cases. The lack of fully developed reuse planning business processes leads to the insufficient exploitation of reuse potentials. Consequently, the utilization of existing management information systems are not suitable for this analysis. Often, there is no explicit information about reuse and reusability; rather, there is a strong dependency on individual implicit knowledge. There are no standardized evaluation methods to assess reusability, especially the incorporated flexibility. The cost evaluations within the reuse projects are based mainly on simple cost or net present value comparisons of the price for a new system and the effort to reuse existing systems. These simple comparisons are not able to model and evaluate the flexibility which is enabled by the equipment reuse. Alternative evaluation methods for this purpose were described in Chap. 3.

In the following, the analysis methodology for two phases during the automotive product development process are described and knowledge-based software systems to support the involved business processes are proposed. In the early planning phase, the system operator/OEM has to access information on possible reuse rates

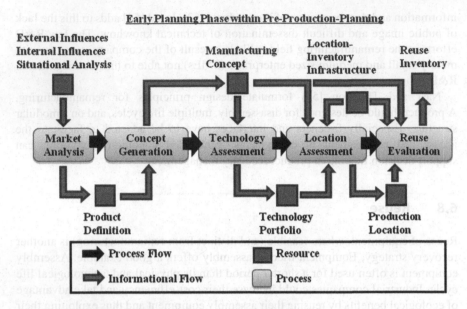

Fig. 6.14 Early planning phase within preproduction planning

for new projects. In the implementation phase, the system integrator has to access the existing system and plan the reuse process of the production system. This early planning phase is illustrated in Fig. 6.14.

During that early product planning phase the planner needs a rough decision on which of the existing equipment is potentially reusable. To do so, three decision categories could be distinguished: reusable, conditionally reusable, and not reusable. To decide which category a particular piece of equipment is in, an evaluation based on the data in the knowledge base will take place. The evaluation categories are divided into economic criteria, technology criteria, and risk criteria. Using a portfolio method, the individual results will be calculated to one overall evaluation. Additionally, the social and ecological impact of reusing the system will be calculated using standard LCA software and available social indicators. The result of these calculations will support the sustainability reporting of enterprises as proposed by the Global Reporting Initiative.

Figure 6.15 shows the processes that the system integrator is involved in. The reusability evaluation contains all processes that are needed to assign particular equipment to a particular process. That includes the evaluation and estimation of the equipment condition, the comparison of this project to older ones, and the assignment of equipment as well as the layout configuration.

The system integrator will be supported in evaluating the economic changeability potential with a real option valuation approach. Matlab® software can be structured for a simplified development of a decision tree and the calculation of the option value is generated by the reuse possibility of the equipment. The process is simplified by assumptions which are generated through the knowledge

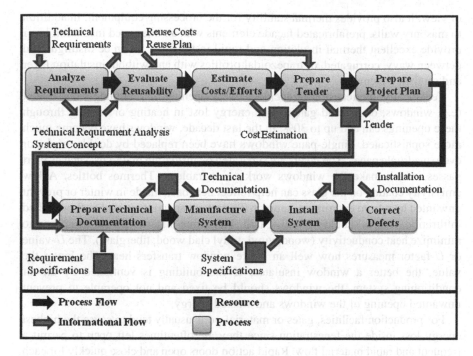

Fig. 6.15 System integrator planning process

base. The software should support the preparation of the purchase order and the generation of a reuse project plan, including required maintenance tasks.

6.9   Approaches for Sustainable Factory Design

6.9.1   Energy-Efficient Building Technology

Energy-efficient building technology should not only be a topic for residential buildings. The term "green factory" often refers to the factory buildings designed and built to minimize energy use and built to recover resources. Industrial buildings often consume and waste a huge amount of energy in terms of electricity and heat. In particular, when a factory or a building gets designed from scratch, energy-efficient building technologies can easily be implemented. However, in existing buildings, small and inexpensive changes can lead to big energy savings and improve the energy efficiency of the whole building. An excellent example of a factory designed to be energy and resource efficient, with respect to building use, is the Ford Motor Company Rouge Plant [67].

The facades of a building have a number of functions including thermal and sound insulation as well as protection from humidity. In particular, thermal insulation for production facilities is not only important from a sustainable point of view; it also provides thermal stability for the processing equipment. In addition to masonry walls, prefabricated façade elements with an integrated insulation layer provide excellent thermal insulation and rapid installation. It can be distinguished between wavy, corrugated, or trapezoidal profiles with and without insulation layer and sandwich elements.

Technically, the fenestration includes every opening in a building's outer shell (e.g. windows, doors, and gates). The energy lost in heating or cooling through these openings can add up to 30%. In the last decade, windows have become much more sophisticated. Single-pane windows have been replaced by double, triple, or even quadruple panes. The layers are separated by insulating materials, such as inert gasses, which make the windows work comparable to Thermos bottles. A low emissivity coating on the glass can help reflect heat back inside in winter or prevent unwanted heat from entering in summer. To prevent an unintentional airflow, called infiltration, frames should be well sealed. The frame material should be chosen to minimize heat conductivity (wood, vinyl, vinyl clad wood, fiberglass). The U-value or U-factor measures how well an entire window transfers heat. The lower the value, the better a window insulates. If the building is ventilated by an air conditioning system, the windows should be fixed and not operable to prevent unwanted opening of the windows and wasted energy.

For production facilities, gates or material ports usually tend to cause the highest energy loss inside the fenestration since they are oftentimes left open to permit a frequent and rapid material flow. Rapid action doors open and close quickly for each crossing and therefore prevent energy losses. This effect can be enhanced if two rapid action doors are arranged consecutively as a lock.

The inside of a building should be illuminated as much as possible with daylight, since this is the most natural and energy-efficient type of light. The minimum intensity of illumination cannot always be achieved only by daylight. Therefore, artificial lighting is oftentimes necessary. In order to reduce energy consumption, a sensor that measures the intensity of illumination can be installed. A controller in connection with a dimmer regulates the artificial light such that the overall intensity of illumination is permanently above the minimum threshold.

In spacious rooms or halls the lighting can be divided into zones such that only used areas are illuminated. If the facility is not used 24 hours a day, then the lighting can be automatically switched off after the last shift or after the last person usually leaves the building. After that, emergency lighting provides a minimum illumination. Motion sensors are another option to control the lighting in the facility. However, frequent switching-on and -off is generally detrimental for the lamps such that motion sensors should not be used in highly frequented areas like toilets and stairways.

6.9.2 Energy-Efficient Process Technology

Inside a factory, water is used for consumption, cooling, fire, and process water. Generally, water is withdrawn from the local water systems. To reduce the consumption, there is the possibility to obtain water from a well or to collect rainwater in a reservoir. The latter is cost-free and has the advantage that the water is already de-ionized and can therefore be used directly in the cooling system. Regular water from the water system would eventually cause fouling in closed networks, like the cooling system, if not manually de-ionized. In cooling towers, where a part of the water evaporates, a large amount of water is needed and can cost-effectively be replaced by well or rainwater. Furthermore, rainwater can be used for other purposes, which do not require clean and drinkable water (e.g., flushing the toilet).

Parts of equipment in a factory produce a lot of heat as a by-product. Usually, this heat is released in the environment and therefore lost. For example, compressors to produce compressed air emit 96% of the energy they consume as heat either in the form of hot air or in the form of hot water. For that reason, compressed air is one of the most expensive and least energy-efficient ways of transporting energy. Other examples of equipment that produce a lot of heat are compression refrigeration or extrusion machines.

The waste heat can be transferred to a cooling water circuit and then be used for heating or even cooling by means of a plate heat exchanger. However in each case, it has to be investigated if the total waste heat is sufficient to justify the investment in a plate heat exchanger and even to replace a boiler.

Cooling that is required in certain processes is usually produced with refrigeration equipment. The latter consumes a lot of energy to cool down the liquid in the cooling circuit. Depending on the local climate (in particular the temperature and humidity), cooling towers can unload the refrigeration machine sometimes up to 100%. The cooler and dryer the air, the more efficient the cooling tower works. It has to be distinguished between dry and hybrid cooling towers. Dry cooling works as long as the temperature outside is below a certain threshold dependent on the temperature level that has to be achieved. Hybrid cooling towers can at first be used in the same way as dry cooling towers. As soon as the temperature exceeds the threshold, the fins in the cooler are sprayed with water in order to take advantage of the evaporative heat loss. In this way, the temperature level that can be achieved in the cooling circuit goes down to the cooling limit temperature.

6.9.3 Compressed Air

Compressed air was referred to in the previous section with respect to energy consumption. In a factory, it is often regarded as the fourth utility, after electricity, natural gas, and water. In manufacturing plants, compressed air is widely used for actuating, cleaning, cooling, and drying parts, as well as removing metal chips [68]. However, the cost of compressed air production is one of the most expensive and least understood processes in a manufacturing facility [69]. The cost of electric power used to operate an air compressor continuously for a year (about 8200 h) is usually greater than the initial price of the equipment [70]. Per million British

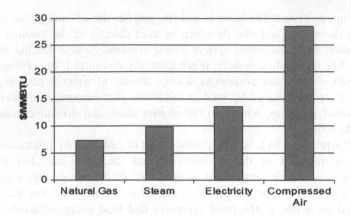

Fig. 6.16 Relative cost of factory utilities

Thermal Unit of energy delivered, compressed air is more expensive than the other three utilities as shown in Fig. 6.16.

Besides cost issues, compressed air production consumes a huge amount of energy. It is estimated that about 3–9% of the total energy consumed in US in 1997 was for air compression in manufacturing [71]. Common supply patterns for compressed air include the following three options:

(1) *Plant air*: the whole plant is supplied with compressed air from the air house with pipelines spread out in the plant to provide compressed air to all facilities.
(2) *Point-of-use*: each machine is exclusively supplied by an independently installed air compressor.
(3) *Local Generation*: a certain number of machines are grouped together and supplied by an air compressor.

An example of compressed air use analysis is given by Yuan et al. [72]. Compressed air is used relatively indiscriminately in automotive manufacturing due to its ease of setup. There is no need for additional maintenance or special machines; the task can be accomplished by adding piping. In addition, as a form of energy, compressed air represents no fire or explosion hazard; as the most natural of substances, it is clean and safe and regarded as totally "green" [73].

Yuan et al. [72] found that plant air was the more typical supply of compressed air, but that it suffers from a number of limitations:

1. *Infrastructure Complexity*: A plant air system is usually very complicated, especially in large-scale use like that of an automotive facility. The complexity of the system not only brings challenges in layout and supply of the compressed air through the pipelines, but it also creates high costs and difficulties in subsequent maintenance and operations.
2. *Low Efficiency*: The plant air system operates at low efficiency. Typically, less than 60% of the total compressed air consumed contributes directly to the

goods and services for which production was intended [74]. Leaks are a major problem in plant air supply. They can occur anywhere along the pipeline, especially at interfaces and connectors of supply lines such as couplings, fittings, filters, regulators, valves, and thread sealants. Eliminating leaks entirely is not feasible, and the system infrastructure often makes ongoing repair costly and difficult.

3. *Energy Storage*: The compressed air system works as a power source to the production facilities. Pressure is difficult to store due to inevitable leakage and limited storage space efficiency.

4. *Excess Supply*: Stable air pressure is critical for the practical operation of production facilities. Further, the supply system must be able to meet maximum demand at all times. These demands require that excess compressed air be provided for the system, which not only consumes more energy than needed but also increases equipment wear and causes reduced equipment life and higher maintenance cost.

5. *High Cost*: With the large amount of compressed air consumed daily for various operations in manufacturing facilities, the total cost may be reduced by using other supply patterns.

An investigation of the compressed air supply methodologies was conducted to determine which of the three supply options mentioned previously could be the most cost-effective considering energy and related operating costs, including equipment purchase and operation.

In this analysis the compressed air was used to supply operating air to 24 computer numerical control (CNC) machine tools in an automotive manufacturing facility. For the "local option" a number of different alternative compressed air supplies were evaluated. Here, local generation is to group a certain number of CNC machines together to be supplied by one air compressor. The advantage of this pattern is elimination of the problematic long, complicated pipelines in the plant air supply pattern, which makes the system easy and convenient to control and maintain. Here, local generation is analyzed with six options, namely LG1 through LG6, according to the number of CNC machines a single compressor is able to supply. Different commercially available compressors were evaluated [72].

An environmental analysis of the competing technologies was also done. The environmental analysis was divided into several stages. First, process identification is the analysis of the demand side of requirements for compressed air in various components of the 24 CNC machines to be served. The step following is the collection of the performance specifications of the supply side of compressed air. These are primarily the compressor characteristics, including energy and load efficiency. Facility information is obtained and utilized to perform a complete cost of ownership (CoO) study. The problem is set up to perform the study for a set of alternative air supplies. The data obtained from various sources were entered into the analysis and the results are presented with a functional unit of cents or kilowatt-hours per 1,000 cubic feet of air to conform to industry standards.

In addition to the environmental analysis, a CoO study was done of the two alternative usage patterns to compare their cost of ownership with that of plant air in

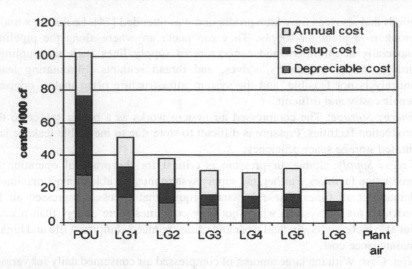

Fig. 6.17 First year cost of supply configurations (with setup cost)

supplying the CNC machines. The costs considered were depreciable costs, setup costs, and annual costs for the operation of the equipment annually. Depreciable costs are equipment cost divided by equipment life; setup costs include those for installation, transportation, and engineer training; annual costs include those for electricity, space consumed on the factory floor, maintenance, and some consumables.

The analysis found that a "local generation" option would be best when compared with POU and plant air options for cost consideration and energy efficiency (Fig. 6.17). Employment of local generation instead of plant air could potentially save $2,000 to $3,200 dollars and 95,000 KWH each year on the CNC milling machines at the automotive production facility. Meanwhile, local generation offers numerous advantages over plant air in regard to reliability, simplicity, leakage prevention, and flexibility. Local generation is supplied by relatively short pipelines, which may lead to a significant reduction of losses due to leaks. Extra local compressors may be connected to the CNC machines in parallel, which automatically builds a great deal of redundancy into the system. Furthermore, the scale of local generation compressors enables greater flexibility as machines and processes change.

6.10 Conclusions

This chapter covered a wide range of topics under the general heading of closed-loop aspects of production systems. Here, this included the life cycle impacts of processes, systems, and facilities in the context of green and sustainable manufacturing, utilization, plant services, and plant design. This can be translated into cost of ownership if the suitable metrics are used to analyze the costs and benefits of alternative approaches—whether it is reuse, remanufacture, or reorganization.

References

1. Emblemsvag J (2003) Life cycle costing: using activity based costing and Monte Carlo simulation to manage future costs and risks. Wiley, New Jersey
2. Harms R, Fleschutz T, Seliger G (2008) Knowledge based approach to assembly system reuse. Proceedings of the ninth biennial ASME conference on engineering systems design and analysis ESDA
3. Woodward D (1997) Life cycle costing—theory, information acquisition and application. Int J Project Manag 15(6):335–344
4. Ferry DJO, Flanagan R (1991) Life cycle costing a radical approach. Construction Industry Research and Information Association, London
5. Jolliet O, Dubreuil D, Gloria T, Hauschild M (2005) Progress in life cycle impact assessment within the UNEP/SETAC life cycle initiative. Int J LCA 10(6):447–448
6. Seliger G (ed) (2007) Sustainability in manufacturing—recovery of resources in product and material cycles. Springer, Berlin
7. Jovane F, Yoshikawa H, Alting L, Boer CR, Westkamper E, Williams D, Tseng M, Seliger G, Paci AM (2008) The incoming global technological and industrial revolution towards competitive sustainable manufacturing. CIRP Ann Manuf Technol 57:641–659
8. Wiendahl HP, ElMaraghy H, Nyhuis P, Zдh M, Wiendahl HH, Duffie N, Brieke M (2007) Changeable manufacturing—classification, design and operation. CIRP Ann Manuf Technol 56:783–809
9. Koren Y (2005) Reconfigurable manufacturing and beyond. CIRP third international conference on reconfigurable manufacturing systems, Keynote paper
10. Dashchenko A (ed) (2005) Reconfigurable manufacturing systems and transformable factories. Springer, Berlin
11. Takata S, Kirnura F, van Houten F, Westkämper E, Shpitalni M, Ceglarek D, Lee J (2004) Maintenance: changing role in life cycle management. CIRP Ann Manuf Technol 53 (2):643–655
12. Fleschutz T, Harms R, Selgier G (2009) Valuation of assembly equipment reuse with real options. proceedings of production and operations mangement society conference (POMS) 20th annual conference, Orlando, Florida
13. Fleschutz T, Harms R, Seliger G, Rusina F, Bottero F (2008) Evaluation of the reconfiguration and reuse of assembly equipment. Proceedings of second cirp conference on assembly technology and systems, University of Windsor, Toronto
14. Fleschutz T (2009) Beruecksichtigung der oekologischen Dimension in Investitionsentscheidungen bei Montageanlagen. Oekobilanzierung 2009: Ansaetze und Weiterentwicklungen zur Operationalisierung von Nachhaltigkeit, KIT, Karlsruhe, Germany, 157–166
15. Wuppertal Institute for Climate, Environment and Energy (2009) Research for sustainable development. http://www.wupperinst.org/uploads/sbs_dl_list/Factor_W.pdf. Accessed 7 Jan 2011
16. PRé Consultants (2011) SimaPro software. http://www.pre.nl/simapro/impact_assessment_methods.htm. Accessed 7 Jan 2011
17. Shimoda M (2002) LCA case of machine tool. Symposium 2002 of the Japan Society for Precision Engineering Spring Annual Meeting, pp 37–41
18. Diaz N, Helu M, Jayanathan S, Chen Y, Horvath A, Dornfeld D (2010) Environmental analysis of milling machine tool use in various manufacturing environments. IEEE international symposium on sustainable systems and technology, Laboratory for Manufacturing and Sustainability, Berkeley
19. Dornfeld D, Lee DE (2007) Precision manufacturing. Springer, New York
20. Jovane F, Alting I., Armillotta A, Eversheim W, Feldmann K, Seliger G, Roth N (1993) A key issue in product life cycle: disassembly and the environment. Ann CIRP 42(2):651–658
21. Harjula T, Rapoza B, Knight WA, Boothroyd B (1996) Design for disassembly and the environment. Ann CIRP 45(1):109–114

22. Ilgin MA, Gupta SM (2010) Environmentally conscious manufacturing and product recovery (ECMPRO): a review of the state of the art. J Environ Manag 91(3):563–591
23. Dhar NR, Kamruzzaman M, Ahmed M (2006) Effect of minimum quantity lubrication (MQL) on tool wear and surface roughness in turning AISI-4340 steel. J Mater Process Technol 172:299–304
24. Weinert K, Inasaki I, Sutherland JW, Wakabayashi T (2004) Dry machining and minimum quantity lubrication. Ann CIRP 53(2):511–537
25. Klocke F, Eisenblatter G (1997) Dry cutting. Ann CIRP 46(2):519–527
26. Munoz AA, Sheng P (1995) An analytical approach for determining the environmental impact of machining processes. J Mater Process Technol 53(3):736–758
27. Krishnan N, Sheng PS (2000) Environmental versus conventional planning for machined components. Ann CIRP 49(1):363–366
28. Narita H, Kawamura H, Norihisa T, Chen LY, Fujimoto H, Hasebe T (2006) Development of prediction system for environmental burden for machine tool operation. JSME Int J Series C 49(4):1188–1195
29. Narita H, Desmira N, Fujimoto H (2008) Environmental burden analysis for machining operation using LCA method. Proceedings of the 41st CIRP conference on manufacturing systems
30. Dahmus JB, Gutowski TG (2004) An environmental analysis of machining. Proceedings of the 2004 ASME international mechanical engineering congress and RD&D expo. Anaheim, California
31. Taniguchi M, Kakinuma Y, Aoyama T, Inasaki I (2006) Influences of downsized design for machine tools on the environmental impact. Proceedings of the MTTRF 2006 annual meeting
32. Gutowski T, Dahmus J, Thiriez A (2006) Electrical energy requirements for manufacturing processes. Proceedings of the 13th CIRP international conference on life cycle engineering, pp 623–627
33. Ashby MF (2009) Materials and the environment: eco-informed material choice. Butterworth-Heinemann, Burlington, MA
34. Kalpakjian S (1996) Manufacturing processes for engineering materials. Prentice Hall, Upper Saddle River, NJ
35. Haapala K, Rivera JL, Sutherland J (2009) Reducing environmental impacts of steel product manufacturing. Trans North Am Manuf Res Instit SME 37:419–426
36. Baniszewski B (2005) An environmental impact analysis of grinding. SB thesis. Dep Mech Eng Massachusetts Institute of Technology, Cambridge, MA
37. Independent System Operators (ISO) New England (2010) Connecticut: 2010 State Profile. http://isonewengland.org/nwsiss/grid_mkts/key_facts/ct_01-2010_profile.pdf. Accessed 11 Oct 2010
38. Fthenakis VM, Kim HC (2007) Greenhouse gas emissions from solar electric and nuclear power: a life-cycle study. Energy Policy 35(4):2549–2557
39. Pacca S, Horvath A (2002) Greenhouse gas emissions from building and operating electric power plants in the upper colorado river basin. Environ Sci Technol 36(14):3194–3200
40. Hondo H (2005) Life cycle GHG emission analysis of power generation systems: Japanese case. Energy 30(11–12):2042–2056
41. Nyberg M (2009) 2008 Net system power report. California Energy Commission. CEC-200-2009-010
42. Gagnon L, Bélanger C, Uchiyama Y (2002) Life-cycle assessment of electricity generation options: the status of research in year 2001. Energy Policy 30(14):1267–1278
43. Diaz N, Helu, M, Jarvis A, Toenissen S, Dornfeld D, Schlosser R (2009) Strategies for minimum energy operation for precision machining. Proceedings of machine tool technologies research foundation 2009 annual meeting
44. Tawakoli T, Hadad MJ, Sadeghi MH, Daneshi A, Stöckert S, Rasifard A (2009) An experimental investigation of the effects of workpiece and grinding parameters on minimum quantity lubrication—MQL grinding. Int J Mach Tools Manuf 49:924–932

45. Wakabayashi T, Inasaki I, Suda S (2006) Tribological action and optimal performance: research activities regarding MQL machining fluids. J Machining Sci Technol Special Issue Environmentally-Conscious Machining 10(1):59–85
46. Sreejith P, Ngoi B (2000) Dry machining: machining of the future. J Mater Process Technol 101:287–291
47. Shefelbine W, Dornfeld DA (2004) The effect of dry machining on burr size. Laboratory for Manufacturing and Sustainability, Berkeley, UC. http://escholarship.org/uc/item/603201b9. Accessed 8 Oct 2010
48. Aoyama T, Kakinuma Y, Yamashita M, Aoki M (2008) Development of a new lean lubrication system for near dry machining process. Ann CIRP 57:125–128
49. Thierry M, Salomon M, van Nunen J, Van Wassenhove LN (1995) Strategic issues in product recovery management. Calif Manag Rev 37(2):114–135
50. Beamon BM (1999) Designing the green supply chain. Logist Inform Manag 12(4):332–342
51. VDI 2243 (2002) Recycling-oriented product development. Beuth, Berlin
52. Ijomah WL, Bennett JP, Pearce J (1999) Remanufacturing: evidence of environmentally conscious business practice in the UK. First international symposium on environmentally conscious design and inverse manufacturing. DOI 10.1109/ECODIM.1999.747607
53. Amezquita T, Hammond R, Salazar M, Bras B (1995) Characterizing the remanufacturability of engineering systems. ASME Design Technical Engineering Conferences, Boston, pp 271–278
54. Lund RT (1996) The remanufacturing industry: hidden giant. Boston University, Boston, MA
55. Nasr N, Thurston M (2006) Remanufacturing: a key enabler to sustainable product systems. Thirteenth CIRP international conference on life cycle engineering, Leuven, Belgium
56. Östlin J (2008) On remanufacturing systems: analysing and managing material flows and remanufacturing processes. Dissertation, Linköpings Universitet
57. Sundin E, Tang O, Marten E (2005) The Swedish remanufacturing industry: an overview of present status and future potential. Twelfth CIRP international conference on life cycle engineering, Grenoble, France
58. Östlin J (2005) Material and process complexity—implications for remanufacturing. Fourth international symposium on environmentally conscious design and inverse manufacturing, Tokyo, Japan
59. Fleischmann M, Krikke HR, Dekker M, Flapper SD (2000) A characterisation of logistics networks for product recovery. Omega Int J Manag Sci 28:653–666
60. Caterpillar US Website (2009) Company website. http://www.cat.com. Accessed 9 Jan 2009
61. Caterpillar (2008) Caterpillar Remanufacturing Singapore Overview. Presentation. http://web.mit.edu/sma/events/career_fair/2008/cat_overview.pdf. Accessed 9 Jan 2009
62. ABB Ltd (2010) Automation division. ABB Certified Refurbished Robot. Available at http://library.abb.com/global/scot/scot241.nsf/veritydisplay/69eb4c57b6bd0c7bc125758a0030f3ca/$File/ABB. Certified Refurbished Robot v4_final.pdf. Accessed 21 April 2009
63. Morel MK (2006) Refurbished robots can save replacement costs. Robotics World 24(1):4–7
64. Steinhilper R (1998) Remanufacturing: The ultimate form of recycling. Fraunhofer IRB Verlag, Stuttgart, Germany
65. Weule H, Buchholz C (2001) Method for the assessment of reuse suitability within modular assembly systems. Assembly Automation 21(3):241–246
66. Schmälzle A (2001) Bewertungssystem für die Generalüberholung von Montageanlagen: Ein Beitrag zur wirtschaftlichen Gestaltung geschlossener Facility-Management-Systeme im Anlagenbau, Dissertation, Universität Karlsruhe
67. Ford Motor Company (2011) Company website. http://media.ford.com/article_display.cfm?article_id=2847. Accessed 7 Jan 2011
68. Sweeney R (2002) Cutting the cost of compressed air. Machine Design 74(21):76
69. Risi JD (1995) Energy savings with compressed air. Energy Eng J Assoc Energy Eng 92(6):49–58

70. Kaya AD, Phelan P, Chau D, Sarac HI (2002) Energy conservation in compressed air systems. Int J Energy Res 26:837–849
71. Curtner KL, O'Neill PJ, Winter D, Bursch P (1997) Simulation-based features of the compressed air system description tool, XCEEDTM. Proc Intl Building Performance Simulation Assoc Conf, Prague, Czech Republic, September 8–10
72. Yuan C, Zhang T, Rangarajan A, Dornfeld D, Ziemba W, Whitnbeck R (2006) A decision-based analysis of compressed air usage patterns in automotive manufacturing. SME J Manuf Syst 25 (4):293–300
73. Cox R (1996) Compressed air—clean energy in a green world. Glass Int 19(2):2
74. Foss RS (2002) Managing compressed air energy part I: demand side issues. Maintenance technology online. http://www.mt-online.com/articles/0801_mngcompressedair.cfm. Accessed 26 May 2006

Semiconductor Manufacturing

7

Sarah Boyd and David Dornfeld

> *It was not so very long ago that people thought that semiconductors were part-time orchestra leaders and microchips were very, very small snack foods.*
>
> Geraldine Ferraro

Abstract

Semiconductor manufacturing, one of the fields of manufacturing in which the USA has played a dominant role for decades, is seen as a major consumer of resources and a source of environmental impact. The objective of this chapter is to introduce the basics of semiconductor manufacturing and, then, look at a detailed analysis of the energy and global warming impact of manufacturing one typical semiconductor product, the complementary metal oxide semiconductor (CMOS) chip. Process steps are reviewed, materials, consumables, and waste streams described, and then an example of applying life-cycle analysis to CMOS fabrication and use (including materials processing through transportation and use phases) is presented. The level of data detail required is illustrated along with trends in manufacturing and environmental impact over several technology nodes.

S. Boyd (✉)
PE INTERNATIONAL, Inc. & Five Winds Strategic Consulting, 344 Boylston St, 3rd Floor, Boston, MA 02116, USA
e-mail: s.boyd@pe-international.com

D. Dornfeld
Laboratory for Manufacturing and Sustainability (LMAS), Department of Mechanical Engineering, University of California at Berkeley, 5100A Etcheverry Hall, Mailstop 1740, Berkeley, CA 94720-1740, USA
e-mail: dornfeld@berkeley.edu

D.A. Dornfeld (ed.), *Green Manufacturing: Fundamentals and Applications*,
DOI 10.1007/978-1-4419-6016-0_7, © Springer Science+Business Media New York 2013

7.1 Overview of Semiconductor Fabrication

The term "microfabrication" is used interchangeably to describe technologies that originate from the microelectronics industry as well as small tool machining for mechanical parts production. This chapter focuses on microfabricated semiconductor devices. These are principally integrated circuits ("microchips"), but similar technologies are used to fabricate a wide variety of other products such as microsensors (e.g., air bag sensors), inkjet nozzles, flat panel displays (FPDs), laser diodes, and so on. Similarly, the term "micromachining" is used for semiconductor processing as well as mechanical machining at the micron scale. Overall, microelectronic fabrication, semiconductor fabrication, MEMS fabrication, and integrated circuit technology are terms used instead of microfabrication, but microfabrication is the broad general term.

Modern semiconductor devices require hundreds of manufacturing process steps, using high purity materials in energy-intensive clean rooms. The high purity requirements of integrated circuit manufacturing extend from the starting material (silicon wafers), throughout the process flow to nearly all of the chemicals and materials used in production and the exacting specifications of the manufacturing clean room environments. Each wafer is processed to form layers of patterns using a repetition of the following three basic processes. First, thin films of conductive, insulating, or semiconductor materials are deposited on the wafer by physical or chemical means. This is followed by a lithography step, in which a pattern is transferred from a mask to a sacrificial photosensitive material. Finally, the thin films are etched through the pattern in the photosensitive material resulting in its transfer to the deposited film. Other processes are related to growing insulating layers (oxidation), introduction and control of dopants used to moderate transistor active regions (ion implant), chemical mechanical planarization (CMP) of films, and wafer cleaning. After processing, each wafer contains hundreds of individual devices called "dies," which are tested, diced, and packaged into chips. Fundamental processes for manufacturing modern silicon wafers are described in detail below. This chapter discusses the manufacture of integrated circuits, though some of the following process steps are also used in manufacturing FPDs, photovoltaics (PVs), and other semiconductor products.

Semiconductor production is highly resource intensive and generates a wide variety of emissions, some of which have global effects. The processes used to manufacture semiconductors emit several major classes of pollutants, including global warming gases (e.g., CF_4, NF_3, C_4F_8), ground level ozone-forming volatile organics (e.g., isopropyl alcohol, formaldehyde), hazardous pollutants (e.g., arsenic, fluorine), and flammable materials (e.g., silane, phosphine). Semiconductor fabrication facilities also consume large volumes of water and energy, and the high purity chemicals used in production are highly refined and thus have high "embodied energy." The upstream environmental effects due to chemicals manufacturing, as well as fabrication facility (fab) infrastructure and equipment, represent significant components of the environmental impact profile of semiconductor manufacturing. The use phase of semiconductor devices results in indirect environmental and human health impacts resulting from energy-related emissions which, in the case of logic devices, has been shown to dominate impacts over the product life-cycle. The end of life of a semiconductor chip results in lead emissions if there is lead present in the chip's leadframe solder. After 2006, the EU's Restriction on Hazardous Substances, commonly known as RoHS, banned the use of lead in electronics and most manufacturers switched to lead-free solders world-wide to comply with this regulation. While other effects from end-of-life disposal of semiconductor devices may exist, they are not included in this discussion because they have never been specifically measured.

This chapter cannot cover this important subject in any real detail. Some examples of challenges and approaches to analysis are given. For a more extensive treatment of this topic readers are referred to [1] and [2].

7.2 Microfabrication Processes

Microfabrication refers to a set of technologies utilized to produce microdevices. Many of the technologies are derived from very different processes and "arts," often not connected to manufacturing in the traditional sense. For example, lithography derives from early printing techniques using etched plates to transfer patterns to paper. Planarization technology, formerly referred to as only polishing, comes from optics manufacturing dating back to the time of early astronomers and physicists. Much of the vacuum techniques also come from nineteenth century physics research. Electroplating is also a nineteenth century technique adapted to produce micrometer scale structures, as are various stamping and embossing techniques.

In the fabrication process for microdevices, a number of types of processes must be performed, in a defined sequence, often repeated many times. In the fabrication of memory chips, over 30 lithography steps, 10 oxidation steps, 20 etching steps, 10 doping steps, etc. are carried out as part of this process. Typical process steps include:

- Photolithography
- Etching (microfabrication), such as RIE (reactive-ion etching) or DRIE (deep - reactive-ion etching)
- Thin film deposition, see, e.g., sputtering, CVD (chemical vapor deposition), evaporation
- Epitaxy
- Thermal oxidation
- Doping by either thermal diffusion or ion implantation bonding
- Chemical mechanical planarization (CMP)
- Wafer cleaning also known as "surface preparation"

The complexity of microfabrication processes can be described using a number of measures, but "mask count" is typical. Mask count refers to the number of different pattern layers that will make up the final microelectronic device. Modern microprocessors are made with upwards of 30 masks while only a few masks may be used for a microfluidic device or a laser diode. The fabrication process is not unlike multiple exposure photography in that many individual patterns (each on a mask) must be aligned with each other in the various layers of the process to create the final structure. In between the stages of fabricating these layers a number of other critical process steps occur (for example, etch/strip and CMP). The masks used in photolithography constitute a major portion of the cost of processing the microdevice and, recently, a number of so-called maskless techniques relying on writing processes without the mask have been discussed [3].

A few of the more prominent process steps are described in more detail below to illustrate the complexity.

7.2.1   Lithography

A major component of semiconductor fabrication is photolithography. The lithography process is the means whereby patterns are transferred onto a substrate (e.g., silicon, gallium arsenide, etc.). The pattern is used to isolate areas for subsequent etching to create trenches for interconnects and lines or to protect the substrate from etching. The patterns are written on glass plates called reticles, much like the glass slides used in earlier forms of photo presentations with projectors. These are the masks referred to in the previous section. Lithography is used because it allows exact control over the shape and size of the features created, and because it can create patterns over an entire surface simultaneously. The main disadvantages are that it is primarily used for creating 2D (i.e., "flat") structures, and, as with other semiconductor processes, requires extremely clean operating conditions. In a complex integrated circuit (for example, CMOS), a wafer will go through the photolithographic cycle up to 50 times. Lithography machines are designed to enhance throughput but necessarily require sophisticated mechanical structures, control, and metrology to maintain pattern quality at high exposure speeds.

Photolithography involves a number of steps in a series of often repeated combinations including:

- Substrate preparation
- Photoresist application
- Soft-baking
- Exposure
- Developing
- Hard-baking
- Etching

and various other chemical treatments (thinning agents, edge-bead removal, etc.) in repeated steps on an initially flat substrate.

A typical cycle of silicon lithography would begin with the deposition of a layer of conductive metal several nanometers thick on the substrate. A layer of photoresist—a chemical that hardens or softens when exposed to light (often ultraviolet)— is applied on top of the metal layer by spinning the substrate under a stream of photoresist. The mask, basically a transparent plate with opaque areas printed on it, is placed between a source of illumination and the wafer, selectively exposing parts of the substrate to light. The photoresist is then developed during which areas of unhardened photoresist undergo a chemical change. After a hard-bake, a series of subsequent chemical treatments etch away the material under the developed photoresist, and then etch away the hardened photoresist, leaving the material exposed in the pattern of the original photomask.

A characteristic of photolithography clean room environments is that the filtered fluorescent lighting contains no ultraviolet or blue light to prevent accidental exposure of the photoresist. Most types of photoresist are available as either "positive" or "negative" resists. With positive resists the area that is opaque (masked) on the photomask corresponds to the area where photoresist will remain upon developing (and hence where conductor will remain at the end of the cycle). Negative resists will create the inverse—any area that is exposed will remain, while any areas that are not exposed will be developed. After developing, the resist is usually hard-baked before subjecting to a chemical etching stage which will remove the metal underneath.

7.2.2 Oxidation and Annealing

Thermal oxidation of silicon produces silicon dioxide, a high-quality insulator that can also be used as a gate oxide, as a stress barrier for nitride (pad oxide), or to prevent contamination in ion implant (screen oxide) and undoped silicate glass (USG) applications (barrier oxide). In this process, silicon on the wafer reacts with oxygen or oxidizing chemicals such as N_2O in a temperature range of 1,173–1,373 K, to form silicon dioxide. Vertical furnaces can be used to produce thicker oxide layers on batches of wafers, or Rapid Thermal Processing (RTP) equipment can be used to produce thinner oxide layers on individual wafers. Annealing is another high temperature process that can be performed in RTP chambers. Annealing is used to control the concentration profile of dopant and to reduce defects.

7.2.3 Wet Cleans

Wafer cleaning is the most frequently occurring category of process step in the production of a wafer due to the need for contamination removal and surface conditioning before sensitive deposition and thermal treatments. Wafers are cleaned after each photoresist (PR) removal and CMP step, and before nearly every oxidation, anneal, and deposition step. A brief summary of the main types of wafer cleaning steps is given in Table 7.1

Table 7.1 Wafer cleaning methods [1]

Clean name and components	Purpose	Usage
SPM or "Piranha" clean H_2SO_4:H_2O_2: UPW (1:1:6)	Removal of organic materials	SPM is used as the primary means of PR removal or to supplement plasma PR strip
SC1 (Standard Clean 1) NH_4OH:H_2O_2: UPW(1:1:6)	Removal of organic and some metal impurities	
SC2 (Standard Clean 2) HCl:H_2O_2:UPW (1:1:6)	Elimination of metallic and alkaline contaminants	
Hydrofluoric oxide strip HF:UPW (1:100)	Silicon dioxide ("oxide") layer removal	Before any metal deposition and as a part of, or after, most dielectric etch steps

7.3 Facility Systems

7.3.1 Resource Use

In addition to the specific chemicals used in processes such as cleaning and etching process steps vast volumes of other fluids, resources, and materials are used.

For example, ultra pure water (UPW) or deionized (DI) water are used for numerous cleaning steps in semiconductor manufacturing as seen in Table 7.1. A typical semiconductor fab uses two million gallons of water or more [4]. Out of that, about three quarters of the water is used for UPW and a single wafer may require 2,000 gallons of UPW. UPW is treated to remove minerals, colloids, and bacteria using reverse osmosis, ion exchange, and/or ultra filtration processes. UPW is expensive both to produce and to treat for release because of its use of water, energy, and consumable materials. However, used rinse UPW is typically much cleaner than municipal supply water and may be treated and reused in the UPW system or reclaimed for other uses in a fab. In certain cases, fabs that reclaim UPW water for cooling towers may eliminate the need to purchase municipal water for their cooling towers altogether.

Nitrogen is an inert gas that is used to purge chambers and pipes between processes. Nitrogen may also be used to condition other purge gasses to desired temperatures. Generally, it is produced onsite and piped into a fab rather than delivered in tanks. Argon, carbon dioxide, or clean dry air (CDA) may also be used for similar purposes. Use of CDA instead of nitrogen may represent significant cost and energy savings for a fab. CDA is likewise produced onsite via filtration and dehumidification. It is used for various purposes but particularly for drying wafers following wet clean steps.

Maintaining a clean room requires numerous energy-intensive components, including fans, filters, air conditioning, and dehumidifying. The air in a clean room must be filtered to remove particles corresponding to the clean room class, recirculated to provide a specified number of air changes per hour, and pressurized with make up air so that contaminants do not enter the room even as air or process gasses are exhausted out of the clean room. The class clean room refers to the max number of particles up to 0.5 μm in size allowed per cubic foot of air. The number of air changes per hour varies from double to triple digits with air grades varying from grade D to grade A (for explanation of grades of clean room see, for example [5]).

Process cooling water (PCW) systems are another energy-intensive component of a semiconductor fab. PCW is used to cool process chambers, pumps, and abatement equipment. PCW is either cooled via the cooling tower alone or chilled via an additional chiller. The energy intensity of PCW may be reduced by moderating the temperature differential between the supply and return PCW.

7.3.2 Abatement

The oldest and most fundamental of the facility abatement systems is the "house scrubber," an enclosed, water-sprayed matrix of inert mesh. This system captures gaseous inorganic emissions, largely acids, which are sent as liquid effluent to the acid waste neutralization (AWN) system, which continuously monitors and corrects the pH of the incoming liquid waste.

Gaseous ammonia is emitted in small quantities from most nitride CVD processes, either as unreacted precursor or as a byproduct emission. Fabs with gaseous ammonia exhaust are fitted with a separate ammonia exhaust system and scrubber in order to prevent particulate formation, clogging, and corrosion in the acid exhaust system. Gaseous ammonia waste is captured using a water scrubber similar in design to the facility acid scrubber but about a tenth of the size.

Complementary metal oxide semiconductor (CMOS) logic fabs use large quantities of both liquid ammonia and sulfuric acid in wafer cleaning processes. Liquid ammonia, collected via drain, may be recycled on site using membrane filtration or distillation, or treated using sulfuric acid to produce ammonium sulfate. In this model, the latter is assumed and thus ammonium sulfate, which results from the neutralization of ammonia and sulfuric acid effluents in the AWN system, is among the liquid wastes produced in the highest volume by wafer fabrication in this model.

There are several combinations of treatment methods that may be used to address the liquid effluent of copper CMP processes. Copper CMP waste treatment is described in the work of Krishnan [6, 7] as a sequence of ion exchange, microfiltration, activated carbon filtering, and filter pressing. An ion exchange resin bed removes copper and is regenerated at the fab using sulfuric acid, to produce $CuSO_4$ liquid waste. Slurry particles are filtered and pressed into a solid non-hazardous waste which is sent to a landfill. The remaining water contains less than 2 ppm dissolved copper and is sent to the AWN system. The concentrated CuSO₄ liquid is sent offsite as hazardous waste to be electrowinned for copper recovery or possibly purified into a useable byproduct.

The fluoride waste system treats fluoride wastewater using $CaOH$ and a flocculant material to produce non-hazardous solids containing calcium fluorite (CaF_2).

CVD steps emitting per-flouro-compounds (PFCs) require combustion and water scrubbing or plasma point-of-use (POU) abatement because water scrubbing alone does not break down these compounds (and in some cases may form reactive fluorinated byproducts). CVD steps emitting silane or hydrogen above flammable concentrations also require immediate combustion of their emissions in POU systems due to the risk of explosion in exhaust lines. Implant processes emitting phosphine and arsine are typically abated using cold bed adsorption systems.

7.4   Green Manufacturing in the Semiconductor Industry: Concepts and Challenges

The semiconductor manufacturing process is exceptional in the large variety of chemistries that it employs. As can be seen in the previous section, wafer processing involves a number of different acidic (the hydrofluoric and sulfuric acids used in wafer cleans), basic (wafer clean steps including ammonia), oxidizing (wafer cleans using peroxide), and other highly reactive chemistries (fluorine used in etching), as well as compounds which are extremely toxic (arsine and phosphine used in implant). The equipment used to administer these reactions must be designed to protect the manufacturing personnel, following safety rules outlined by government agencies such as OSHA and standards (e.g., SEMI S2) developed within industry groups such as Sematech.

As all mainstream semiconductor manufacturing equipment currently sold and used follows these regulations, the direct human health impacts and risks within the fab have been nearly eliminated in normal operation. (Though, hazards still exist in cases of catastrophic breakdown, fire, or earthquake.) Once these chemicals leave the equipment, they must be further handled and neutralized by the POU and facility abatement systems, in a safe and efficient way. While the guidelines and standards for equipment safety are enough to thoroughly guide and ensure the design of safe equipment, the design and operation of facility abatement is a much more complex undertaking. The abatement and neutralization of emissions is not as predictably efficient or controlled as the reaction of chemicals within the process equipment in part because the processes used to neutralize emissions to the extent necessary to make them safe for release into the environment do not need to be as precise as those used within the process chamber. Additionally, within the facility abatement systems (the house gaseous waste, fluorine abatement, and AWN systems), the chemistry of the combined emissions of the many processes running on site can be unpredictable. Facility abatement systems are designed to continuously measure the incoming waste stream and adjust the neutralization chemistry accordingly.

Nevertheless, neutralization of an unpredictable waste stream cannot be as efficient or controlled as that of a known waste stream.

When facility abatement systems are not operating ideally, or were not originally designed or built to sufficiently handle the current waste streams entering them, a variety of environmental impacts can result. For example, the "house scrubber" (facility gaseous abatement system) may be accepting significant concentrations of gaseous fluorine (F_2), either because no POU abatement is set up on plasma etching equipment or because POU systems are not sufficiently scrubbing the F_2 gas. This gaseous fluorine will react with water to a small extent to form OF_2, a reactive and highly toxic gas [8]. Another product of the reaction of fluorine with water is HF. When fluorinated compounds are effectively abated from processes at POU, the resulting liquid HF is sent to a fluorine waste treatment system which is separate from the house AWN system. Any HF captured in the house scrubber system could not be effectively treated before being released into the environment, as it would already be mixed in with the larger volume of non-hazardous waste. Ineffective abatement of fluorine and the consequent release of reactive fluorine species into the environment could result in human health and ecological impacts.

While the potential environmental and health impacts from semiconductor manufacturing are understood and, in most cases, successful efforts are made to eliminate or mitigate them, the global warming potential (GWP) impacts associated with certain PFCs were not recognized or controlled until many years after the introduction of their use.

PFCs are an important group of emissions from semiconductor manufacturing due to their high infrared absorption, long lifetimes, and consequential global impact. These compounds are used in wafer etching and include CF_4, C_2F_6, NF_3, and SF_6. For this reason, global warming impacts are an important impact category to consider in the production of ICs.

The abatement of some PFC emissions are regulated by the Kyoto Protocol (in Annex I and II nations) and, in 1999, the World Semiconductor Council (WSC), which includes the semiconductor industry associations of Japan, Europe, Korea, Taiwan, and the USA, issued a position paper which committed members to PFC emissions reduction by 10% of 1995 or 1999 baseline levels by the end of 2010. The China Semiconductor Industry Association (CSIA) joined the WSC in 2006 but did not sign on to the climate protection agreement at that time. In 2009, CSIA stated an intended plan to join the WSC agreement on PFCs but did not commit to a baseline year for that goal, and has not yet as of the year 2010 [9].

Although these two agreements have resulted in tremendous progress in the reduction of semiconductor PFC emissions, more than half of semiconductor production occurs outside of Kyoto Protocol Annex I and II nations, and, in 2008, almost 20% of semiconductor production capacity was held in China, Singapore, and Malaysia, where the industrial consortia have not committed to the WSC PFC goals. Semiconductor capacity has continued to grow in those countries where PFC emissions control is not required by any public agreement or national policy. NF_3 is not regulated by the Kyoto Protocol, but is among the PFCs which are used in highest volume in the semiconductor industry [10, 11].

7.5   Use-Phase Issues with Semiconductors

Most consumer products which consume more than a few watts in operation will have a use phase which dominates energy consumption impacts among all of the life-cycle stages. Depending on the electricity mix in the location of use, therefore, the use phase of an integrated circuit will in most cases be the largest contributor among its life-cycle stages to primary energy and water use, as well as GWP, acidification, ground-level ozone formation, and other impacts related to electricity generation. Even in the case of an IC with a low power consumption, the use phase is an important contributing stage to life-cycle impacts.

Most ICs are built into products which have a use phase which dominates energy consumption impacts among all of the life-cycle stages. The importance of the use phase has thus been no secret to those concerned with the environmental impacts of integrated circuits and computers in general. Indeed, the longest running and likely most well-known environmental initiative concerning electronics is Energy Star [12]. Energy Star is a labeling program operated in cooperation between the US Environmental Protection Agency and Department of Energy. The Energy Star program develops testing protocols, collects data, and sets thresholds for the definition of energy efficiency in a variety of categories of consumer electronics. Manufacturers who certify their product to the standard may then print the Energy Star label on their product or packaging. Computers and monitors became the first consumer products to carry the Energy Star label in 1995.

In addition to the improvement in efficiency related to hardware (e.g., lower power consumption in integrated circuit chips, more efficient battery technologies), power management through software applications has also played a significant role in improving the efficiency of laptops and other electronic products. For example, advanced, or "dynamic," operating system-integrated chip power management allows software to shut down the central processing unit (CPU) when the user is inactive. Introduced in the early 2000s, this software was available as part of both the Windows 2000 and XP operating systems, but neither of these platforms had advanced power management settings enabled by default, and the functionality was often not enabled by the user. In 2007, when many used their computers with advanced power management features disabled due to the default Windows settings, a market research study found that as many as 60% of computer users in the USA did not shut down their computer at the end of the day. These computers which were left on (termed "zombie" computers), which would otherwise have been put to sleep via advanced power management, resulted in the needless emission of an estimated 14 million tons of CO_2 that year (Alliance to [13]). In Windows 7, the default settings for shipment were for lower power consumption, which supported wider use.

The Climate Savers Computing initiative is a more recent industry initiative concerning the use phase of computing which put a particular focus on software-integrated power management. Climate Savers has served as a platform for collaboration and technical standard-setting to improve efficiency in hardware as well as increase the adoption and consumer use of advanced power management.

More recently, other manufacturers and distributors have been working on labeling standards for indicating the consumption of consumer electronics. The Sustainability Consortium has an Electronics Sector Working Group that is focused on creating scientifically grounded and transparent metrics for measuring and reporting environmental and social impacts of electronics [14]. A large group of companies and organization representing the sector is working to create these metrics using a life-cycle based platform and will be in an ISO-certifiable and index-ready format.

7.6 Example of Analysis of Semiconductor Manufacturing

7.6.1 Introduction

The semiconductor manufacturing process is complex and, unsurprisingly, determination of the associated emissions and their impacts is not a straightforward task. Nevertheless, it is instructive to look at a detailed analysis of the life-cycle energy and global warming emissions of CMOS logic. This section is taken substantially from the doctoral thesis of one of the chapter authors, Sarah Boyd, and can be found in its entirety in [1] or [2].

Information and communication technology (ICT) has the potential to reduce the impact of human activities on the environment. In order to fully understand the environmental benefits of ICT, the life-cycle impacts of computer systems must be compared with those of the products and services they replace. The questions of whether reading news on a handheld device rather than newspaper or purchasing books from an online retailer instead of from a bookstore reduces environmental impact are two examples of this sort of comparison in the recent literature [15–17]. While, initially, the replacement of traditional products such as newspapers by a small fractional increase in the use of a handheld mobile device seems a winning environmental trade-off, there has been increasing concern over the large energy demands of the Internet infrastructure, with data center energy demand in the USA reaching 1.5% of the national total in 2006 and estimates of 2011 demand surpassing 10 billion kWh [18].

Among the numerous parts which compose the IT infrastructure, semiconductor chips are among the most resource intensive to produce as well as the most difficult to characterize for the purposes of life-cycle assessment (LCA). While it may be possible to estimate the environmental impacts of a cable or plastic computer housing knowing only their masses and material types, the impacts associated with a semiconductor chip are not represented well by the substance of the device itself. While a logic chip may weigh only a few grams, the chemicals and water required to produce it weigh many kilograms. There is a need for a more detailed and transparent life-cycle inventory (LCI) for semiconductor products.

CMOS is the dominant device structure for digital logic. The CPU in desktops, laptops, handheld devices, and servers, as well as nearly all embedded logic (the chips in appliances and toys) are CMOS-based. Every 1–3 years, a new generation or technology node of CMOS is introduced, based on design laws which have been established through industrial collaboration. Due to the cooperation necessary to plan and achieve the goals for each generation, there is considerable homogeneity among the devices manufactured by the major logic producers at each technology node. A generic version of CMOS may thus be used to represent logic products from many different manufacturers.

This section provides a summary of a life-cycle energy analysis for CMOS chips over seven technology generations with the purpose of comparing energy demand and GWP impacts of the life-cycle stages, examining trends in these impacts over time and evaluating their sensitivity to data uncertainty and changes in production metrics such as yield. Chips of generic CMOS logic, produced at a semiconductor fabrication facility (fab) located in Santa Clara, California are evaluated at each technology node over a 15-year period, from the 350 nm node (circa 1995) to the 45 nm node (circa 2010). This study is composed of production-related LCA data, based on emission measurements, process formulas, and equipment electrical tests, combined with previously published LCA data for chemicals, electricity, and water, as well as publicly available use-phase data for computer chips. A hybrid LCI model is used. Wafer production, electricity generation, water supply, and certain materials are represented by process LCA data, while the remaining materials are described using economic input–output life-cycle assessment (EIO-LCA) methods [19]. While life-cycle energy and GWP of emissions have increased on the basis of a wafer or die as the functional unit, these impacts have been reducing per unit of computational power. Sensitivity analysis of the model shows that impacts have the highest relative sensitivity to wafer yield, line yield, and die size and largest absolute sensitivity to the use-phase power demand of the chip.

The methodology used in the study is detailed first, including the materials and other resource and production data sources covering the full range from material production to end of life of the CMOS device. The results of the study are then summarized. This example illustrates the level of detail necessary to adequately measure, or estimate, the impact in terms of energy use and GWP of such a complex production.

7.6.2 Methodology

The scope of this LCA includes materials production, wafer processing, die packaging, transportation, and use of the logic chip, Fig. 7.1. The LCA model is hybrid, using a combination of process-based LCA and economic input–output (EIO) LCA data (Table 7.1). The functional unit is one packaged die, but in order to allow further analysis and to investigate trends, results are also presented per wafer and per million instructions per second (MIPS). The stages of analysis cover from materials production through end of life.

At end-of-life, it is assumed that there is no recoverable energy value in the chip. Other end-of-life impacts are not included because the functional unit of this LCA is the chip alone and past studies of electronic waste impacts have generally

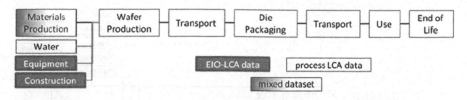

Fig. 7.1 Life-cycle stages with data source types [1]

concerned the computer as a whole. A great deal of effort has been focused on the end-of-life of computer systems because irresponsible recycling practices can produce dramatic and visible human health and environmental impacts. The major pollutants associated with e-waste (flame retardants, polychlorinated biphenyls, dioxins/furans, polycyclic aromatic hydrocarbons, lead, cadmium, and mercury) are largely emitted from the incineration or chemical breakdown of circuit boards, wiring, housing, and displays. Although there may be harmful emissions from the decomposition or combustion of a logic chip, these have not yet been measured in isolation, but remain an important topic for future work. Because there is no positive energy value and no global warming impacts at end-of-life, the net impact in this life-cycle stage is zero.

In order to clarify the model structure and in order to demonstrate the sensitivity of results to variation in model parameters, the inventory model is described algebraically. The contributors to the life-cycle energy requirements (e_{total}) and global warming potential (GWP) of life-cycle emissions (g_{total}) are illustrated in (7.1) and (7.2).

$$e_{total} = e_{up} + e_{inf} + e_{prod} + e_{trans} + e_{use} \tag{7.1}$$

e_{up}: energy for upstream materials;
e_{inf}: energy for infrastructure;
e_{prod}: energy for production;
e_{trans}: energy for transportation;
e_{use}: use-phase energy.

$$g_{total} = g_{up} + g_{inf} + g_{prod} + g_{trans} + g_{use} \tag{7.2}$$

g_{up}: GWP of emissions due to upstream materials;
g_{inf}: GWP of emissions due to infrastructure;
g_{prod}: GWP of emissions due to production;
g_{trans}: GWP of emissions due to transportation;
g_{use}: GWP of emissions due to use-phase energy.

A schematic of the mass and energy flows used in this analysis relative to the fab is shown in Fig. 7.2, from [1].

The compositions of these various terms from (7.1) and (7.2) are now described.

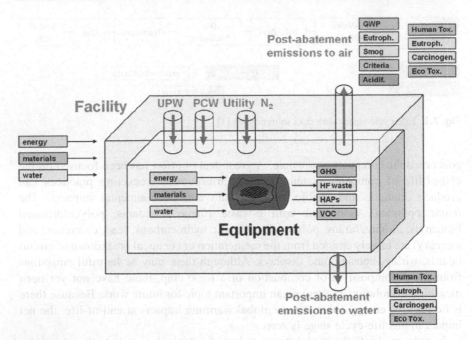

Fig. 7.2 Overview of mass and energy flows considered in the fab model [1]

7.6.3 "Upstream" Materials

7.6.3.1 Chemicals

Among the life-cycle impacts of semiconductor products, the importance of energy-related emissions from the production of high purity chemicals has been noted previously by a number of authors [20–23]. The limited LCA data available for exotic and/or high purity semiconductor process chemicals remains a challenge in quantifying these impacts. The formulas for advanced semiconductor processing materials such as CMP slurries are closely held intellectual property. Chemical textbooks and handbooks simply do not contain information about the production processes used to make them, and it is challenging to identify the dominant production method among patent filings, as enterprises will at times file multiple patents describing different production pathways or describe production recipes broadly. While LCA data are available for some basic chemicals used in wafer manufacturing, such as elemental gases, metals, and common acids, it is usually representative of the industrial grade, with a purity of 99% or lower, rather than ultra-high purity or semiconductor grade (99.9997–99.9999999% pure).

This example uses a method of LCA data collection by which data based on process descriptions are used where available, and data from the Carnegie Mellon EIO-LCA database are used where costs are known. When no process LCA data and no cost information are known, an estimate for the energy intensity of chemical manufacturing developed by Overcash is used [24]. In this study, the

Table 7.2 Summary of data sources [1]

Silicon	Process LCA
Chemicals	Process and EIO-LCA
Infrastructure and equipment	EIO-LCA
Fabrication	Process LCA
Electricity	Process and EIO-LCA
Water	Process and EIO-LCA
Transportation	Process LCA
Use	Process LCA

"pharmaceuticals and medicines" rather than "photographic film and chemicals" commodity sector (NAICS #325400) is used in the EIO analysis for those materials which are high value specialty chemicals (those with a purchase price over $1,000 per kg), since the economic value of these materials is represented more closely by the former sector. The organic chemicals (NAICS #325190) and inorganic chemicals (NAICS #325180) commodities are used for the remaining materials, as appropriate. Although additional impact categories are available for those materials analyzed using EIO-LCA, the inventory is limited to primary energy demand and the GWP of emissions. Data sources for all inventory materials are detailed in [1]. The uncertainty of EIO-LCA data is given as one order of magnitude for each result. The uncertainty of process data from textbooks and manuals is assumed to be zero, because it is unknown but assumed to be small as compared with other chemical LCA data sources. All data sources and impact values for materials using published process energy data are given in [1].

7.6.3.2 Silicon
Silicon is the purest substance used among all semiconductor process materials. There are several processing steps that raw silica takes to become a pure silicon wafer, the substrate of semiconductor devices. Raw silica is refined into metallurgical grade silicon, which is twice refined to produce a single crystal ingot that is then sliced into wafers. The high embedded energy of the final product (approx. 2,000 kg) is due not only to the energy intensity of these processes but also to a cumulative low yield caused by the losses at each step. Full descriptions of the energy requirements and environmental emissions of high purity silicon production are available from previous sources [21, 25]. The LCA data provided by Williams and used in this study [21] is duplicated here for clarity, Table 7.2.

7.6.3.3 Water
Since the focus of this example is the production of chips of generic CMOS logic at a fab located in Santa Clara, California, the environmental impacts associated with the Santa Clara water supply are modeled. Modeling is done using information from the Santa Clara Valley Water District and previous work on LCA of California water supplies by [26]. The Santa Clara Valley Water District infrastructure is composed of three treatment plants for local and imported water, one recycled

Table 7.3 Energy intensity of silicon production [21]

Process step	Electrical energy/kg Si out (kWh)	Si yield (%)
Refining Silica to Mg-Si	13	90
Mg-Si to trichlorosilane	50	90
Trichlorosilane to polysilicon	250	42
Crystallization of polysilicon to sc-Si ingot	250	50
Sawing Sc-Si ingot to Si wafer	240	56
Process chain from silica to wafers	2,127	9.5

Table 7.4 Global warming intensity of Santa Clara Water [1]

	Local supply	Imported	Recycled
Contribution of source (%)	45	51	4
kWh/l	0.0021	0.0019	0.0002

water treatment facility, 142 miles of pipelines, and three pumping stations. According to a report from the district board, approximately 51% of the water used in Santa Clara is imported, while 45% comes from local sources and the remaining 4% from recycled stocks [27]. Most water imported to Santa Clara comes from the Sacramento-San Joaquin River Delta via the South Bay Aqueduct, though a small fraction also comes from the Hetch-Hetchy reservoir via the San Francisco water system. Local water sources include groundwater basins and ten surface reservoirs. The life-cycle environmental impacts evaluated by Stokes for imported and recycled water from the Oceanside Water District in San Diego are applied, on a per volume basis, to the imported and recycled fractions of water in the Santa Clara system. Life-cycle environmental impacts associated with Santa Clara's locally sourced water are estimated based on the energy required for treatment and distribution of imported water in Stokes' model of Marin's water treatment works. The global warming emissions intensity for the power utility in Santa Clara (Pacific Gas and Electric), 280 g CO_2eq./kWh, is used. The energy intensity and percent contribution of each source is presented in Table 7.3. The resulting global warming emissions per liter of water provided in Santa Clara is 0.6 g CO_2eq.

7.6.3.4 Infrastructure and Equipment

The energy use and GWP for infrastructure and equipment are evaluated using EIO-LCA. Rock's Law (which says that the cost of a semiconductor chip fabrication plant doubles every 4 years) is used to estimate the total cost of the fabrication facility and the costs of wafer fabrication equipment are taken as 70% of the total cost of the fab, based on a commonly stated approximation. Expenditures are depreciated over a 10-year period, using a straight line schedule, yielding an annual cost which is corrected to 1997 dollar values using the average US inflation rate over the 1995–2008 period of 2.7%. Total costs for the building and equipment for each technology node are provided in Table 7.4.

Table 7.5 Cost of fab infrastructure and equipment [1]

Year	1995	1998	1999	2001	2004	2007	2010
Technology node	350	250	180	130	90	65	45
Equip. cost, depreciated ($M/year)	42	71	84	119	200	336	400
Construction cost, depreciated ($M/year)	18	21	25	30	36	43	51

7.6.3.5 Electricity

The emissions associated with electricity use at the different geographical locations of each life-cycle stage are reflected in the model. In the fabrication and use stages, emissions factors for electricity are specific to California, while the stages of chemical and infrastructure production are represented by each US industry average GWP emissions factors, via EIO-LCA [19].

The environmental impacts associated with electricity supplied to the California plant are evaluated using two previous LCA of electricity generation, data from the EPA and information from Santa Clara's electric utility, Pacific Gas and Electric. The electricity mix of Pacific Gas and Electric in 2008 was 47% natural gas, 23% nuclear, 13% large-scale hydroelectric, 4% coal, 4% biomass or other waste combustion, 4% geothermal, 3% small-scale hydroelectric, 2% wind, and 0.1% solar photovoltaic [28]. The life-cycle GHG emission factors (g CO_2eq./kWh) for natural gas, coal, large-scale hydroelectric, and solar photovoltaic power are taken from the work of [29], while that for nuclear electricity is taken from a study by [30]. Direct GHG emissions for geothermal and biomass combustion are taken from the EPA [31]. Small hydro is considered to have the same impacts as large hydro. A national average for the Chinese grid of 877 g CO_2eq/kWh, based on a previous LCA [32], is used for the production scenario in China.

In order to facilitate comparison with preceding studies, for most life-cycle stages, the convention of 10.7 MJ of primary energy per kWh electricity is used. This represents a worldwide average value for fuel consumption in electricity production [21]. The primary energy intensity of electricity supplied in Santa Clara is not documented, and since there have been no studies which provide net fuel intensity of nuclear, geothermal, wind, or the other non-combustion generation technologies used by the California grid, the fuel intensity of the electricity used in fabrication is taken as the worldwide average. In actuality, the primary energy intensity of Santa Clara electricity is estimated as the world average. Since most of the thermal generation in California is combined cycle natural gas combustion, and the contribution of renewables and nuclear are higher than the world average, the net primary energy demand for electricity production is somewhat lower than 10.7 MJ/kWh. For the purposes of this example, however, the global average is used.

The fuel intensity of electricity in China, however, is higher, with an average value of 12 MJ/kWh of electricity, due to an average lower conversion efficiency of power plants as well as higher losses in transmission and distribution [32].

7.6.3.6 Semiconductor Manufacturing

In this analysis the primary model for wafer manufacturing is located in Santa Clara, California, USA. A separate scenario for production in China is developed in order to demonstrate the environmental effects of using China's electricity supply mix and neglecting PFC abatement. Although PFC emissions may be abated in some fabs in China, the assumption is made that there are few if any controls on PFC emissions at the Chinese production site.

The mass and material flows are accounted at the level of the fab and equipment, see Fig. 7.2.

A complete summary of changes to the process flow for each device over the technology nodes covered in this example is given in [1]. The process change which has allowed the greatest reduction in GWP from one technology node to the next is the switch from in situ plasma generation to remote plasma generation for etch and post-dielectric deposition chamber cleaning.

Facility and Process Equipment Energy Demand. While device design, process complexity, and the length of the process flow grow continuously, total fab energy consumption has not increased at the same pace and has at times decreased in the past decade due in large part to facility efficiency improvements. These changes are reflected in the model; at each technology node, improvements are made to certain facility equipment, such as the water chillers or exhaust pumps, which allow reduced energy consumption. Rising energy costs as well as pressure to achieve GHG emission reduction goals set by the World Semiconductor Council have driven fabs to reduce their total energy consumption. These efforts are reflected in the industry goals set in the ITRS, which show an ongoing effort to reduce facility energy consumption on a kWh/cm^2 wafer area basis. The trend may be verified using an EIO perspective. By normalizing per unit of silicon area used, rather than by economic value of production, energy consumption can be analyzed independent from increases in off-shoring and outsourcing of fabrication by US companies or the increasing economic value of products. US Census data from 1995 to 2005 show that the total electricity consumed by the semiconductor industry in the USA, when normalized per area of silicon consumed by the industry, did not increase significantly from 1995 to 2005 [33, 34]. The energy consumption per area of silicon consumed increases and decreases slightly over time, but was roughly the same in 2005 as in 1995, approximately 1.5 kWh/cm^2 [21, 34].

Energy efficiency goals have largely been achieved through changes to fab facility systems. Throughout the industry, improvements have been made to the energy efficiency of nearly all of the major fab systems: water cooling, exhaust flow, water distribution, clean room air flow, CDA and facility nitrogen delivery systems, and chamber vacuum pumps. Facility energy efficiency improvements can be classified as advancements in both the technologies and in the techniques applied in fab design and operation. Higher efficiency pumps and fans, variable speed drives (VSDs), and improvements in ducting and clean room airflow arrangement such as mini-environments represent technological developments. Reduction of pressures in CDA and exhaust systems, optimization of clean room temperature and air speed, and the use of larger cooling towers to allow reduced chiller size are examples of operational improvements.

These advancements are reflected in the model for each technology node in this study. At the 250 nm node, the pressure maintained in the CDA delivery system is increased to support stepper systems required for this generation's photolithography tools. (This change does not enhance energy efficiency but was necessary to enable pneumatic stepping for lithography.) At the 180 nm node, the air change-over rate (ACR) is reduced in the clean room heating ventilation and air conditioning (HVAC) system, allowing fans speed to be lowered, the scrubber exhaust pumps are upgraded, a smaller and more efficient chiller, using a VSD, is installed; chiller use is also reduced by increasing the size of the cooling towers. Total facility energy consumption is cross-verified against industry reports and published literature [35, 36].

The wafer yield (good chips per wafer), line yield (finished wafers per wafer starts), and chip size are key variables which influence the environmental impacts per chip. The values for these parameters at each technology node are based on industry average data [35].

Power data for process tools are based on measurements taken using three phase power measurement equipment, which have a maximum error of $\pm 2.6\%$. Power requirements for facility systems are determined using mass flow analysis and facility energy consumption models, which are developed based on data from industry and technical reports [37, 38]. Power and facilities requirements for process tools are from process equipment measurements [39] and requirements for abatement equipment are based on manufacturers' specifications, which have an undefined error.

Process Emissions. The abatement of some PFC emissions are regulated by the Kyoto Protocol (in Annex I and II nations) and, in 1999, the World Semiconductor Council (WSC), which includes the semiconductor industry associations of Japan, Europe, Korea, Taiwan, and the USA, issued a position paper which committed members to PFC emissions reduction by 10% of 1995 or 1999 baseline levels by the end of 2010. However, more than half of semiconductor production occurs outside of Kyoto Protocol Annex I and II nations, and, in 2008, almost 20% of semiconductor production capacity was held in China, Singapore, and Malaysia, where the industrial consortia have not joined in the WSC. Thus, although PFC emissions may be abated in some fabs in China, the assumption is made that there are no controls on PFC emissions at the Chinese production site.

GWG emissions from each process step have been determined, pre- and post-abatement, using in situ mass spectrometry and Fourier transform infrared (FTIR) spectroscopy analysis by a procedure which requires mass balance to be closed within 10% of chamber inputs. Each of these measurements thus has a maximum uncertainty of $\pm 10\%$ for each element. For most materials, the uncertainty of the total mass of emissions per finished wafer can be considered as a uniform distribution with variance equal to $(10\%)^2$ of the expected value. For NF_3 which is at more than 30 points during processing of a single wafer the uncertainty is reduced via the central limit theorem, and the total mass flow is modeled as a normal distribution with variance equal to $(3.3\%)^2$ of the expected value. GWPs are taken from [40].

Table 7.6 GWP intensity of transportation [1]

	Distance, fab. to assembly (miles)	Distance, assembly to use (miles)	CO_2 intensity (g CO_2eq./ton-mile)	Energy intensity (MJ/ton-mile)
Truck	50	200	187	2.7
Air freight	3,000	3,000	18	0.38

7.6.3.7 Transportation

Chips are typically cut and packaged at a facility separate from the wafer fabrication site, often in a different country or on a separate continent altogether. Semiconductor products therefore travel twice within the production phase: wafers are transported from the fab to an assembly plant, where they are cut into die, packaged into chips, and tested and finished chips are then transported to the place of eventual use.

The global industry of semiconductor packaging and testing, or "back-end" processing, is clustered in Vietnam, Malaysia, Costa Rica, Puerto Rico, China, and the Philippines. Costa Rica is the closest location to Santa Clara and is therefore the location of assembly designated in this study.

Travel from the wafer fab to the assembly facility is taken as 50 miles by truck and 3,000 miles by plane, and from assembly to the final POU, travel is 3,000 miles by air and 200 miles by truck. Energy consumption and GWP of emissions for truck and air freight are from [41]. The distance of each travel leg and its corresponding GWP impact and energy intensity is given in Table 7.5.

It is assumed that between wafer production and assembly, the finished wafer is transported in a wafer carrier and additional casing with a total weight of 500 g per 200 mm wafer or 700 g per 300 mm wafer. Between assembly and use, the product and packaging has an assumed weight of 20 g regardless of technology node. The total energy and GWP intensity of transport for each technology node is detailed in [1].

7.6.3.8 Use Phase

The use phase represents the power consumption of the chip assuming an average power supply efficiency of 70%. The lifetime of the chip is taken to be 6,000 h (3 years, being used 8 h a day, 5 days per week, and 50 weeks per year) in a 70% active state, representing a business user. An assumption of 3 years is consistent with the literature, which identifies the typical lifespan of personal computers as 2–3 years in business applications and 4–5 years in residential use [33, 42, 43]. The lifetime assumed in this study would also be equivalent to an 18-month lifespan of a data center processor, operating continuously, with 95% uptime, at a 30% activity rate.

The average power requirements for logic chips are taken from the 2001 to 2007 International Semiconductor Manufacturing Roadmap reports [4, 35, 44] and, for years previous, from manufacturer's specifications. These power values represent operation at full capacity or at a 100% activity rate (Table 7.7).

Table 7.7 Use-phase power by technology node [1]

Technology node (nm)	350	250	180	130	90	65	45
Year	1995	1998	1999	2001	2004	2007	2008
Power (W)	14	23	25	61	84	104	146

The average chip power demand has risen from 14 to over 140 W over the past 15 years. The steady increase in power requirements for logic chips is the main cause of rising energy-related life-cycle impacts, as will be shown in Sect. 7.6.4.

In order to compare impacts on a common basis of operational performance, the rate of instructions performed, usually denoted in million instructions per second, MIPS is used, rather than clock speed or transistor density, as a common metric of computational capacity. Transistor density is not ideal as a computational power metric because while increased transistor density usually results in increased computational power, the relation is not necessarily proportional. Although clock speed, which is dependent on transistor density, is used as a popular measure of a CPU performance, computational power is determined by the CPU's architecture, instruction set, cache size, and memory speed as well as clock rate. The rate of instructions in MIPS accounts for both the speed and design of the chip but remains highly dependent on the instruction sequences used to define the metric. Though instruction rate falls short of providing a perfect description of a CPU's performance as processors with different instruction sets or architectures are not comparable, instruction rate is a more representative metric than clock rate or transistor density and is a commonly reported measure of performance. MIPS is thereby used in this analysis as a metric for comparison based on computational performance.

7.6.4 Results and Discussion

As technology has progressed, life-cycle energy use and greenhouse gas emissions have in general been increasing per wafer and per die but decreasing when normalized by computational power. Figure 7.3 shows how total life-cycle energy demands per wafer, per die, and per 1,000 MIPS have changed over the period under study.

The increases in per-wafer and per-die life-cycle impacts have one dominant cause: the escalation of use-phase chip power. The growth in per-wafer impacts, however, is also due to the lengthening of the manufacturing process flow and concomitant expansion in manufacturing infrastructure and equipment, as shown in Fig. 7.4.

At each technology node, the complexity of device design has increased, and the number of process steps required to produce a finished wafer has escalated. In this model, for example, production of a finished wafer entails 147 process steps at the 350 nm node, while the process flow for a 45 nm device consists of a total of 251 process steps. The lengthening of the process flow follows from increasingly

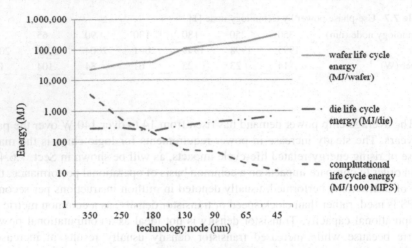

Fig. 7.3 Energy use per die, per wafer, and per 1,000 MIPS by technology node [1]

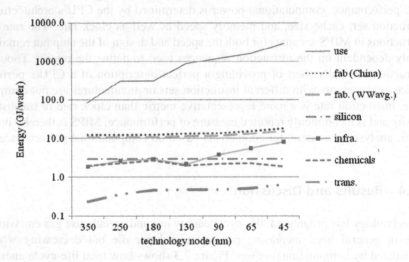

Fig. 7.4 Energy use per 300 mm wafer equivalent, by life-cycle stage, over seven technology nodes [1]

detailed construction necessary to scale down the device's transistors as well as additional interconnect layers to wire them together.

Growth in manufacturing and materials-related impacts over time has been counteracted by shrinking die sizes, which allow more die to fit on each wafer. Thus, use-phase power is the lone reason for increases in impacts per die. For all technology generations, the use phase represents the largest proportion of energy-related impacts per die among the life-cycle phases. The dominance of the use phase has also increased over time, with use contributing about 51% of life-cycle

GWP consumption per die at the 350 nm node, and over 95% per die at the 45 nm node. Despite the long distances that semiconductor wafers and chips are typically shipped during production and prior to use GWP of transportation is almost insignificant due to the small mass of the product.

The improvement of several production performance metrics has allowed reductions in the manufacturing energy and GWP per chip. Line yield reflects wasted processing used for process monitoring, testing, and wafer loss in the form of damage or breakage. Although wafer damage has remained the same over the years, at about 2%, the number of test or monitor wafers per finished wafer has been reduced over the last decade, resulting in higher average line yields [20, 45–47]. Although reduced feature sizes have made maintaining wafer yield difficult, industry reports indicate that wafer yields for full-scale production have not fallen with decreasing device dimensions. Mature wafer yield is assumed to be 75% for all technology nodes, based on ITRS reports [4, 35, 44].

The results of this study enable LCA practitioners to answer important questions concerning the energy-related environmental impacts of computing with greater certainty than ever before. The life-cycle impacts for energy and GWP of semiconductor chips presented in this analysis are more complete, accurate, and transparent than those of any previous study, and data are presented for chips spanning many generations, from 1995 to 2010. Energy and GWP impacts for semiconductor logic chips are clearly dominated by the use phase. Chip power demand and the GWP of use-phase electricity are thus the variables with the largest influence over energy-related life-cycle impacts. Production yield, die size, geographical location or electrical energy supply of the plant, and the choice to abate PFCs are the most important metrics and decisions to be made concerning energy and GWP impacts in the production stage.

7.7 Conclusion

Semiconductor manufacturing, a manufacturing sector which supports growth in many areas of the world economy, has been a major consumer of resources and source of environmental impacts. The industry has made great strides to improve efficiency of resource utilization (from energy to materials) motivated by both cost savings and environmental, health, and safety considerations. Hence, it is almost a "poster child" example of green manufacturing analysis and technology development. This chapter has introduced the basics of semiconductor manufacturing and, then, provided a detailed analysis of the energy and global warming impact of manufacturing one typical semiconductor product, the CMOS chip. This chip, ubiquitous in many consumer products today, offers an example of the benefits of technology advances as well as the challenges. The life-cycle analysis of CMOS fabrication and use (including materials processing through transportation and use phases) illustrated the level of data detail required and demonstrated trends in manufacturing over several technology nodes. This same approach, relating environmental impact in terms of GWP, resource consumption, or other measures, is applicable to a wide variety of products. The LCA study presented in this chapter highlights the importance of identifying of the proper functional unit in semiconductor LCA, as impacts per die, per wafer, and per 1,000 MIPS may be applicable in different contexts, but yield contrasting trends over time.

References

1. Boyd S (2009) Life-cycle assessment of semiconductors. Ph.D. thesis, University of California, Berkeley, Mechanical Engineering
2. Boyd S (2011) Life-cycle assessment of semiconductors. Springer, New York
3. Zhang TW, Bates SW, Dornfeld DA (2007) Operational energy use of plasmonic imaging lithography. Proceedings of the IEEE international symposium on electronics and environment (ISEE), May 2007
4. ITRS (2008) Sematech. The international technology roadmap for semiconductors, 2007 edn. Technical report
5. Filterair (2011) http://www.filterair.info/articles/article.cfm/ArticleID/6ECA4042-A748-400F-8A2FC1798F6B337C/Page/1. Accessed 4 May 2011
6. N. Krishnan, S. Thurwachter, T. Francis, and P. Sheng (2000) The environmental value systems (env-s) analysis: application to CMP effluent treatment options. Proceedings of the Electrochemical Society (ECS) Conference on improving environmental performance of wafer manufacturing processes. ECS
7. Krishnan N (2003) Design for environment (DfE) in semiconductor manufacturing. Ph.D. thesis, Mechanical Engineering, University of California at Berkeley
8. Sherer J (2005) J. Michael Sherer, Semiconductor industry: wafer fab exhaust management, CRC Press, Boca Raton, Florida
9. WSC (2010) Joint Statement of the 14th Meeting of the World Semiconductor Council (WSC), WSC, May 27, 2010, Seoul, South Korea
10. Prather MJ, Hsu J (2008) NF₃, the greenhouse gas missing from Kyoto. Geophys Res Lett 35: L12810
11. Hoag H (2008) The missing greenhouse gas. Nat Rep Clim Change. doi:10.1038
12. Energy Star (2011) http://www.energystar.gov/. Accessed 2 May 2011
13. Alliance to Save Energy (2007) PC energy report 2007, United States. Unpublished report. available online: http://www.climatesaverscomputing.org/docs/Energy_Report_US.pdf
14. Sustainability Consortium (2011) http://www.sustainabilityconsortium.org/electronics/. Accessed 2 May 2011
15. Toffel MW, Horvath A (2004) Environmental implications of wireless technologies: news delivery and business meetings. Environ Sci Technol 38(11):2961–2970
16. Kim J, Xu M, Kahhat R, Allenby B, Williams E (2009) Designing and assessing a sustainable networked delivery (SND) system: hybrid business-to-consumer book delivery case study. Environ Sci Technol 43(1):181–187
17. Rosenblum J, Horvath A, Hendrickson C (2000) Environmental implications of service industries. Environ Sci Technol 34(22):4669–4676
18. EPA (2007) EPA report to congress on server and data center energy efficiency. Technical report, U.S. Environmental Protection Agency
19. Carnegie-Mellon (2009) Carnegie Mellon University Green Design Institute. Economic Input–output life cycle assessment (EIO-LCA), US 1997 industry benchmark model, economic input–output life cycle assessment (EIO-LCA), US 1997 Industry Benchmark model. http://www.eiolca.net. Accessed 15 May 2009

20. Watanabe A, Kobayashi T, Egi T, Yoshida T (1999) Continuous and independent monitor wafer reduction in DRAM fab. In IEEE International symposium on semiconductor manufacturing conference proceedings, IEEE, pp 303–306
21. Williams ED, Ayres RU, Heller M (2002) The 1.7 kilogram microchip: energy and material use in the production of semiconductor devices. Environ Sci Technol 36(24):5504–5510
22. Plepys A (2004) The environmental impacts of electronics: going beyond the walls of semiconductor fabs. Proceedings of the IEEE international symposium on electronics and the environment, IEEE, pp 159–165
23. Krishnan N, Williams E, Boyd S (2008a) Case studies in energy use to realize ultra-high purities in semiconductor manufacturing. Proceedings of the IEEE international symposium on electronics and the environment, IEEE
24. Kim S, Overcash M (2003) Energy in chemical manufacturing processes: gate-to-gate information for life cycle assessment. J Chem Technol Biotechnol 78(8):995–1005
25. Phylipsen GJM, Alsema EA (1995) Environmental life-cycle assessment of multicrystalline silicon solar cell modules (report no. 95057). Technical report, Department of Science, Technology and Society, Utrecht University, The Netherlands
26. Stokes J, Horvath A (2006) Life cycle energy assessment of alternative water supply systems. Int J Life Cycle Assess 11(5):335–343
27. Williams SM (2007) The Santa Clara Valley Water District Fiscal Year 2007–08 five-year capital improvement program. Technical report, Santa Clara Valley Water District
28. PG&E (2008) Power content label. Available online: www.pge.com. Last accessed December 2008
29. Pacca S, Horvath A (2002) Greenhouse gas emissions from building and operating electric power plants in the Upper Colorado River Basin. Environ Sci Technol 36(14):3194–3200
30. Fthenakis V, Kim H (2007) Greenhouse-gas emissions from solar electric and nuclear power: a life-cycle study. Energy Policy 35:2549–2557
31. EPA (2008) The emissions & generation resource integrated database for 2007 (eGrid2007) Technical Support Document. U.S. Environmental Protection Agency, September 2008
32. Di X, Nie Z, Yuan B, Zuo T (2007) Life cycle inventory for electricity generation in China. Int J Life Cycle Assess 12(4):217–224
33. Williams E (2004) Energy intensity of computer manufacturing: hybrid assessment combining process and economic input–output methods. Environ Sci Technol 38(22):6166–6174
34. Census Bureau (2005) Statistics for industry groups and industries, annual survey of manufacturers. U.S. Census Bureau, Washington, DC
35. Sematech (1999) The international technology roadmap for semiconductors: 2005, 2003, 2001 and 1999 editions. Technical report, 1999–2005
36. Hu SC, Chuah YK (2003) Power consumption of semiconductor fabs in Taiwan. Energy 28 (8):895–907
37. O'Halloran M (2002) Fab utility cost values for cost of ownership (CoO) calculations, Technology Transfer #02034260A-TR, available on-line: www.sematech.org/docubase/abstracts/4260atr.htm. Technical Report 02034260A-TR, International Sematech
38. Lawrence Berkeley National Laboratory (2001) High-tech Buildings - Market Transformation Projects: A Final Report to CIEE/PG&E, Technical Report LBNL-49112, Berkeley, California. Available online: http://ateam.lbl.gov/cleanroom/index.htm
39. Krishnan N, Boyd S, Somani A, Raoux S, Clark D, Dornfeld D (2008) A hybrid life cycle inventory of nano-scale semiconductor manufacturing. Environ Sci Technol 42(8):3069–3075
40. IPCC (2007) PCC Climate change 2007: the physical science basis, contribution of working group I to the fourth assessment report of the intergovernmental panel on climate change. Cambridge University Press, Cambridge
41. Facanha C, Horvath A (2007) Evaluation of life cycle air emission factors of freight transportation. Environ Sci Technol 41(20):7138–7144

42. Cremer C, Eichhammer W, Friedewald M, Georgie P, Rieth-Hoerst S, Schlomann B, Zoche P, Aebischer B, Huser A (2003) Energy consumption of information and communication technology in Germany up to 2010. Technical report
43. Roth KW, Goldstein F, Kleinman J (2002) Energy consumption by office and telecommunications equipment in commercial buildings. volume i: energy consumption baseline, adl 72895–00. Technical report
44. ITRS (2007) Sematech. The international technology roadmap for semiconductors, 2007 edition. Technical report
45. Yung-Cheng JC, Cheng F (2005) Application development of virtual metrology in semiconductor industry. Proceedings of the IEEE industrial electronics society conference. IEEE
46. Hsu J, Lo H, Pan C, Chen Y, Hsieh T (2004) Test wafer control system in 300 mm fab. Semiconductor manufacturing technology workshop proceedings, IEEE, pp 33–36
47. Foster B, Meyersdor D, Padillo J, Brenner R (1998) Simulation of test wafer consumption in a semiconductor facility. IEEE/SEMI advanced semiconductor manufacturing conference and workshop, IEEE, pp 298–302
48. IPCC (2001) Good practice guidance and uncertainty management in national greenhouse gas inventories. IPCC-TSU NGGIP, Japan

Environmental Implications of Nano-manufacturing

Chris Yuan and Teresa Zhang

There's Plenty of Room at the Bottom

Richard P. Feynman

Abstract

Green manufacturing for nanotechnologies must be considered, while it is most impactful to do so, during early development and early application stage. There are still a lot of challenges in applying conventional green theories and methodologies to nano-manufacturing technologies, particularly for their overall sustainability assessment and improvement. This chapter provides a basic overview of potential environmental impacts associated with nanotechnology and its manufacturing processes. The fundamental knowledge and scientific methods useful in understanding the environmental impacts of nanotechnology from a holistic view are discussed. A few examples on environmental studies of nano-manufacturing technologies are provided as well. A systematic view of the potential environmental impacts of nano-manufacturing will be helpful for guiding future research in this subject.

8.1 Introduction

Nanotechnology is the science of understanding and control of matter at dimensions of roughly 1–100 nm [1]. Nano-manufacturing is using nanotechnologies to make engineered nano-materials, structures, and devices for novel industrial and social

C. Yuan (✉)
University of Wisconsin, Milwaukee, WI, USA
e-mail: cyuan@uwm.edu

T. Zhang
University of California, Berkeley, Berkeley, CA 94720
e-mail: zhangtw@cal.berkeley.edu

D.A. Dornfeld (ed.), *Green Manufacturing: Fundamentals and Applications*,
DOI 10.1007/978-1-4419-6016-0_8, © Springer Science+Business Media New York 2013

applications. It has been widely believed that nanotechnology is the right technology to address challenges the human beings is and will be facing in industrial, social, and environmental aspects. In the past decade, nanotechnology has undergone rapid development in both R&D and commercialization. Until end of 2009, global spending on nanotechnologies has been approximated as $50 billion in a total [2]. There were a wide variety of nano-materials being developed and produced for commercial use through various nano-manufacturing processes and technologies, such as carbon nanotubes, nano-TiO_2, nano-Ag, etc. The statistical data shows that there have already been more than 1,000 nanotechnology-made-products available on the commercial market, including micro-processors of computers, wax and polish of vehicles, sunscreens and cosmetics of personal cares, and many others [3].

Nanotechnology is relevant because of its high potential in revolutionizing technologies for almost all industrial sectors. Nanotechnology can produce nano-materials through manipulation of atoms and molecules in nanometer scale. The engineered nano-materials have strikingly different properties from their macroscopic or bulk counterparts [4, 5]. It has been understood that at the nanoscale, quantum mechanics, and wave physics versus classical mechanics are jointly functional in the material property determination [6]. As a result, superior material in such aspects including physical, chemical, optical, electrical, catalytic, magnetic, mechanical, and biological properties can be obtained on nano-structured materials which are potential for a wide range of applications on electronics, automobiles, cosmetics, energy storage, medical devices, etc. Use of such materials can lead to significant improvement of products' performance, upgrading of product functionalities, creation of novel applications, development of new technology solutions, etc.

Although nanotechnology is favored due to its great potential in future industrial and social applications, serious concerns have also been raised regarding its potential impact on the environment and eco-system, particularly during its large-scale industrial applications in future. At industrial scale, the environmental impact of nanotechnology will be generated from all the aspects of its applications, ranging from material production, manufacturing, product use, until final disposal.

In general, the types of impacts nanotechnology generates on the environment and eco-system can be divided into two categories: conventional impacts similar to those generated by existing technologies from energy and material consumptions, and unconventional impacts resulting from nano-structured material production and consumption. Within the conventional impact category, the energy consumption and material wastes generated in nano-manufacturing are particularly important in determining its environmental performance. The energy intensity of nano-manufacturing is super higher when compared with conventional manufacturing technologies. Besides, toxic chemicals are widely and heavily used in nanoscale manufacturing for various operations including etching, cleaning, reacting, catalyzing, etc. The consumed energy and toxic materials plus generated nano-process wastes from the nano-manufacturing could have significant impact on the environment and public health.

However, when compared with those conventional environmental impacts, those unconventional impacts resulting from nano-structured materials on human health and eco-system are more concerned by industry and society. The nano-structured materials are so "tiny" in their size and shapes that such materials can easily transport in the environmental media including air, water, and soil. Following different exposure routes as inhalation, ingestion, and dermal uptake, nano-materials can readily enter human body and transport between human cells and might cause unrecoverable health damages.

A comprehensive understanding of the environmental implications of nanotechnologies requires systematic investigation and study of the environmental impact associated with every stage of the application of the nanotechnology. Particularly, green manufacturing is needed to control and minimize the environmental impact of nanotechnologies within the whole manufacturing system. For this purpose, the environmental and human health impact of conventional emissions and nano-particles both need to be assessed. As stated in the first chapter, the best opportunity for environmental impact control and mitigation of nanotechnology is to apply pollution prevention strategies during its early development and application stage, to understand its environmental performance and find technical solutions to minimize its environmental impact. With the environmental performance being scientifically characterized and assessed at the early development stage, the sustainability performance of such emerging manufacturing technologies and their future production systems could be improved and then designed with a minimum amount of emission and waste output.

However, the scientific understanding and investigation of the environmental impact of nanotechnologies are very limited in current stage. Currently, only a limited number of studies have been performed on environmental or sustainability performance of a few nano-materials or nanotechnologies. The studies available in current stage are all focused on the analysis and characterization of those conventional impacts from energy or material consumptions. Overall sustainability studies of nanotechnologies are still at conceptual analysis stage or early discussion stage.

Green manufacturing for nanotechnologies is very necessary, particularly at its early development and application stage. The results from green manufacturing research conducted in the early stage of nanotechnologies would be valuable for use in the future during the environmental management of nano-manufacturing and in the risk control of nano-material emissions in the interests of public health and the environment. But implementation of green manufacturing needs to understand the specific characteristics of the nanotechnology first. There are still a lot of challenges in applying conventional green theories and methodologies on nano-manufacturing technologies, particularly for their overall sustainability assessment and improvement.

This chapter provides a basic overview of potential environmental impacts associated with nanotechnology and its manufacturing processes. In that context, fundamental knowledge and scientific methods for understanding the environmental impacts of nanotechnology from a holistic view is provided. Specifically, the focus is on examining the environmental impacts of nano-manufacturing processes due to their primary role in linking the material flow and energy flows between nano-material production and nano-product use as well as the final disposal.

A few examples on environmental studies of nano-manufacturing technologies are provided as well. A systematic view of the potential environmental impacts of nano-manufacturing will be helpful for guiding future research in this subject.

8.2   Nano-manufacturing Technologies

Before the environmental impacts of nano-manufacturing technologies are presented, a brief summary of nano-manufacturing technologies available in current research practice and industrial production are reviewed. An overview of these nano-manufacturing technologies and their operational characteristics would be helpful in understanding, mitigating and controlling the environmental impacts of these nanotechnologies.

In current practice, there are quite a number of technologies being used for nanoscale manufacturing. Generally, the nanoscale manufacturing technologies can be categorized into the following three groups: top-down, bottom-up, and hybrid nano-manufacturing technologies [7]. A detailed overview of nano-manufacturing methods toward sustainable nano-products is presented by Sengül et al. [8].

8.2.1   Top-down Nano-manufacturing

Top-down nano-manufacturing produces nanoscale features and patterns by using carving or grinding methods including lithography, etching, milling, etc. [8]. Top-down nano-manufacturing is the technology more commonly used today in nano-fabrication due to its high manufacturing efficiency. But top-down nano-manufacturing technologies require very high energy consumption in the high-precision manufacturing of nanoscale features and structures.

With the advanced research and development in precision engineering and lithography technologies, and the relative ease of maintaining control over the precise process and ambient conditions, top-down manufacturing is the more preferred method when compared to bottom-up technology [9, 10].

8.2.2   Bottom-Up Nano-manufacturing

In contrast to top-down nano-manufacturing technologies, the bottom-up nano-manufacturing produces nanoscale features and patterns through atomic-scale manipulation of molecules through chemical reactions or physical processes. The representing bottom-up nano-manufacturing technologies includes chemical vapor deposition (CVD), atomic layer deposition (ALD), physical vapor deposition (PVD), etc. It is generally believed that bottom-up nano-manufacturing are less waste-producing than top-down nano-manufacturing [8]. It is also suggested that bottom-up nano-manufacturing methods should be the ultimate tools for green manufacturing since they allow for the customized design of reactions and processes at the molecular level, thereby minimizing unwanted waste [11].

8.2.3 Hybrid Top-Down and Bottom-Up Nano-manufacturing

Hybrid top-down and bottom-up nano-manufacturing combines the best aspects of top-down and bottom-up nano-manufacturing techniques for massively parallel integration of heterogeneous nano-components into higher-order structures and devices [7]. The hybrid top-down and bottom-up nano-manufacturing technologies are still under development in current stage, but it is expected that the hybrid nano-manufacturing techniques could lead to a major advancement in developing high-precision nanoscale features and simultaneously improving the throughput of nano-manufacturing processes. The hybrid nano-manufacturing could also be promising in offering a solution to environmental issues of nano-manufacturing with a reduced energy consumption and waste generations.

8.3 Conventional Environmental Impact of Nano-manufacturing

Nano-manufacturing technologies, similar to those conventional manufacturing technologies, consume both materials and energy in producing nanoscale features and structures. But these nanoscale operations usually consume much more energy and materials, and accordingly generate more wastes and emissions, than conventional manufacturing on an equal mass basis, due to the fact that nano-manufacturing requires highly precise processes and technologies to manipulate molecules and atoms to make nanoscale materials and devices. The energy consumption and material consumption as well as emissions generated from the consumed energy and materials are classified as conventional environmental impacts of nano-manufacturing, which could be measured and characterized by using conventional analytical metrics as discussed in Chap. 3. In this section, we discuss the conventional environmental impact of nano-manufacturing in the following two aspects: energy use and material consumption.

8.3.1 Energy Use of Nano-manufacturing

Nanoscale manufacturing is very energy-intensive in terms of material processing and nano-feature manufacturing. As more than 70% of world energy is supplied by fossil fuels [12], the large amount of energy consumed in nano-manufacturing would contribute, either directly or indirectly, to generations of a wide range of

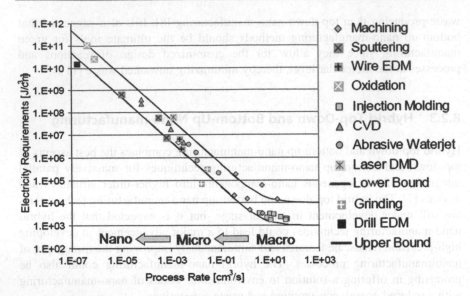

Fig. 8.1 [13] Significant energy use in nanoscale manufacturing. Copyright T. Gutowski, 2006, used by permission

emissions including such greenhouse gases as CO_2, CH_4, N_2O, etc., and such pollutants as SO_x, NO_x, PM, etc. In real practice, understanding the energy consumption and energy flows within a nano-manufacturing process is important to reduce the energy consumption and overall environmental impact of nano-manufacturing.

Although the energy consumption of conventional manufacturing industries is enormous due to their large production volume and product size, the energy intensities of nanoscale manufacturing processes are even higher than those of conventional manufacturing processes. Gutowski et al. have studied the energy consumption of eleven manufacturing processes and found that the energy intensity of nanoscale manufacturing processes (sputtering, CVD, oxidation, etc.) are typically 3–5 orders of magnitude higher than those of conventional manufacturing processes (machining, injection molding, grinding, etc.) [13] (Fig. 8.1).

The high energy intensity of nano-manufacturing technologies directly leads to a high energy requirement on nano-material productions. For example, Kushnir and Sanden studied the energy requirements of carbon nano-particle production and found that the carbon nano-particles are highly energy-intensive materials, on the order of 2–100 times more energy-intensive than aluminum, even with idealized production models [14]. The energy-induced environmental impact, if not successfully reduced during the early development stages of these nano-manufacturing processes, might generate even more significant environmental problems during their future large-scale industrial productions, when compared with those conventional manufacturing industry.

8.3.2 Examples of Energy Analysis in Nano-manufacturing

It is instructive to review two detailed examples to demonstrate the high energy consumption of nano-manufacturing. As mentioned earlier, there are few studies available in environmental analysis of nanotechnologies. Among them, two studies performed at UC-Berkeley evaluated the energy intensity of: Plasmonic Imaging Lithography (PIL), a top-down nanoscale manufacturing technology, and Atomic Layer Deposition (ALD), a bottom-up nano-manufacturing technology.

The first example focuses on PIL. PIL is a new photolithography exposure technique which allows researchers to surpass the diffraction limit of light using the special properties of surface waves or plasmons [15, 16]. The effervescent nature of plasmons requires the use of a near-field scanning system, in this case, one based on hard disk drive technology. PIL has the potential to pattern sub-wavelength features with the precision and throughput time suggested by established hard disk technology.

The second example comes from studies on Atomic Layer Deposition. ALD is a promising nanotechnology process currently under development for a wide range of potential applications. ALD is a bottom-up nanoscale manufacturing technology, derived from Chemical Vapor Deposition (CVD), for depositing highly uniform and conformal thin films by alternating exposures of a surface to vapors of two or more chemical reactants [17]. ALD is different from CVD in that the ALD reactions are separated by complete purging in between, and accordingly, the reaction only takes place on the surface. The ALD surface reaction is self-limited and the film thickness could be accurately controlled in atomic scale. ALD is a cyclic nano-manufacturing technology. An ALD surface reaction is called a cycle and the time to finish a cycle is called cycle time. As needed, the ALD cycling process can be repeated hundreds or even thousands of times to obtain a film layer for a specific thickness. ALD can deposit highly uniform and conformal thin films on extremely complex surfaces [18], and accordingly, has potential applications on a wide variety of products including CMOS chips, flat panel displays, optical filters, medical devices, etc. [19].

In the following, detailed examples are given on energy consumption analysis and improvement of two nano-manufacturing technologies: PIL and ALD.

8.3.2.1 Energy Use of Plasmonic Imaging Lithography

The first example deals with a developing nano-manufacturing process, PIL, on its operational energy use. The direct electrical energy use of the PIL process flow is interesting compared to that of masked optical projection lithography (OPL) and maskless electron beam lithography (EBL).

As shown in Fig. 8.2, the process steps affected by PIL in addition to the exposure step itself are evaluated. Spin coating is included, for example, because PIL uses a hard phase-change resist to facilitate reliable light of the scanning head, and this resist must be sputtered, rather than spin-coated, onto the wafer. Reflecting the capabilities of PIL, we consider the exposure of one layer of a 4 in. wafer with 120 nm line width, 30% density features for each of the lithography techniques.

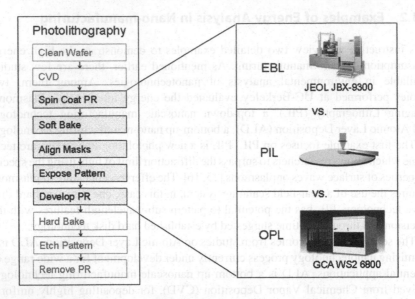

Fig. 8.2 Scope of analysis comparing electron beam lithography (EBL), plasmonic imaging lithography (PIL), and optical projection lithography (OPL)

Energy use data is collected from various sources. Electrical energy use is ideally measured directly using power monitoring equipment with logging capabilities, such as those made by Summit Technology, Fluke, and Dent Instruments. However, it is often not possible to take functional equipment off-line for power measurement. Instead, power supply requirements are collected from physical power supply inspection, and user and installation manuals. Power supply requirements are related to actual power use as follows

$$P_{use} = S \times PF \times UF \times \sqrt{\phi} \qquad (8.1)$$

where P_{use} is the actual power use in kilowatts (kW), S is the apparent power supply in kilovolt-amperes (kVA), PF is the power factor, UF is the ratio of power use to power supply, and ϕ is the phase of the power system (either single or three-phase).

Power factor (PF) is the ratio of real power (P) to apparent power (S), where apparent power is the product of voltage and current [20]. Power factor is 0 for a purely reactive load and 1 for a purely resistive load. California utilities charge for power factors less than 0.85 [21], so a power factor of 0.85 is assumed. The analysis also assumes a usage factor (UF) of 0.67 for all process tools [22]. Process time information is collected from expert machine operators, technicians, and sales representatives based on average batch sizes and processing times. In the case of electron beam lithography exposure, expert operators reported an enormous range of processing times, ranging from 1 h/cm² [23] to over 20,000 h/cm² [24]. We instead calculated operational processing time as

$$t = \frac{D \times A}{I} \qquad (8.2)$$

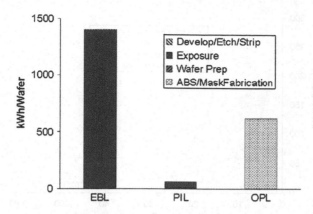

Fig. 8.3 Operational energy use per 4 in. wafer for a one-off design, including the processing energy of an air bearing slider (ABS) for plasmonic imaging lithography (PIL) or a mask for optical projection lithography (OPL)

where t is the time in seconds, D is the dose in C/cm^2, A is the exposure area in cm^2, and I is the current in nA. As with all process tools, startup and idle times are included where appropriate.

Actual power use, processing time, and resulting energy use is shown in Fig. 8.3 for the three lithography techniques.

Figure 8.4 shows energy use as a function of design agility, which we associate with wafers per design change. The crossover point at which OPL becomes more energy efficient than PIL is 11, 39, and 85 wafers for 1, 10, and 100 plasmonic lenses per air bearing slider, respectively. This suggests that PIL is most appropriate for prototyping or highly flexible manufacturing.

These results are also useful for informing the design of the PIL process for energy efficiency. Design for environment is especially valuable to consider at this early stage in the development process, as the design space is still relatively unconstrained.

The range in which PIL is the best option in terms of operational energy use can be expanded most significantly by addressing the biggest sources of energy use in the PIL process flow. The ion laser used in PIL wafer exposure is a good example, consuming 78% of energy used in wafer processing. It draws 10 kW of electrical power, and though it is capable of much higher powered laser beams, it is only used to produce a 260 mW laser beam, representing a loss of four orders of magnitude. Another change that could expand the niche of PIL considerably would be to increase the number of lenses on each air bearing slider. Assuming the use of the same laser, the number of lenses on an air bearing slider head is inversely proportional to wafer exposure time and resulting energy use (Fig. 8.4).

Outside of wafer exposure, the greatest contributor to operational energy use for PIL is sputtering of the resist. To protect the phase-change resist, a metal adhesion layer and a diamond-like coating are also applied to each wafer, requiring two separate sputtering processes. The analysis presented here adds to the practical argument for a liquid photoresist.

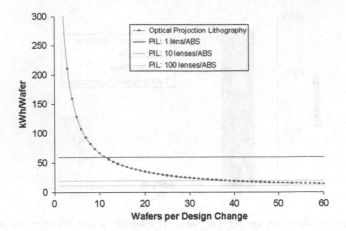

Fig. 8.4 Energy use as a function of design agility

A realistic manufacturing scenario lies somewhere between these bounding cases. While an air bearing slider head is assumed to have a functional life of one wafer pass, a photolithography mask can be used well past the desired life of a particular design.

8.3.2.2 Energy Consumption of Atomic Layer Deposition Nanotechnology

Energy consumption of ALD is much concerned for both economic and environmental justifications. ALD typically has to be operated at the vapor phase of precursor materials and requires ultra-high vacuum conditions for system operations. Energy consumed in the ALD system is mainly for heating, pumping, and process control.

ALD process temperature is a critical parameter in evaluating the energy consumption of an ALD system. In ALD operations, sufficient energy must be supplied into the ALD system to overcome the energetic barrier of the surface chemistry and maintain the reaction conditions.

The study is performed on ALD of Al_2O_3 high-k dielectric films which uses trimethylaluminum (TMA) as the metal source and water as the oxidant

$$2Al(CH_3)_3 + 3H_2O = Al_2O_3 + 6CH_4$$

This is a process developed for replacing conventional SiO_2 dielectric gate in the Metal Oxide Semiconductor Field Effect Transistors (MOSFETs) as the semiconductor process is down-sizing from 90 to 45 nm [25–27]. From experiments, the highest growth rate of Al_2O_3 films was found around 200°C [28]. In this analysis, the process temperature is set at 200°C to ensure a high deposition efficiency of

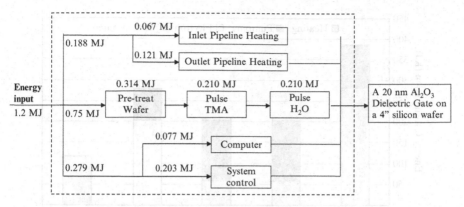

Fig. 8.5   ALD process energy flows on 200 cycles at 4 s at 200°C

Al_2O_3 films. The inlet pipelines are heated to 70°C, while the outlet pipelines are heated to 160 °C to facilitate the pulsing and purging of precursor materials.

During operation, the whole ALD system needs to be heated to maintain the vapor phase of each precursor reactant, and meanwhile, to supply the activation energy to the chemical reaction. In order to have a full understanding of the process energy consumption within the ALD system, a detailed energy flow analysis has been conducted on the ALD of a 20 nm Al_2O_3 high-k dielectric gate at a 4-s cycle time condition. The energy consumption is systematically investigated for each process and component of the ALD system, with an aim to provide useful information for both economic and environmental performance analysis of the Al_2O_3 ALD technology. There are a number of processes and components in the ALD system consuming energy through such activities like heating, pumping, actuating, monitoring, etc.

Power demands of each ALD subprocess and component are measured by a power meter for the Cambridge NanoTech Savannah 200 system used in the experiment. The ALD operations are divided into four categories: pre-treat wafer and system, pulse Al_2O_3, pulse H_2O, system heating and control.

For such a deposition of 20 nm on a 4 in. silicon wafer, the ALD processes consume a total of 1.2 MJ energy. From the perspective of ALD operations, 62% of the total energy is consumed by ALD process operations, 15% consumed by pipeline heating, and 23% used for system control and data acquisition. A detailed energy flow of the ALD process is demonstrated in Fig. 8.5.

The energy flow analysis demonstrates that heating of the outlet pipeline consumes about twice the energy as the inlet pipeline, while the energy consumed in pulsing TMA and H_2O are at the same level. Wafer pretreatment, due to the length of time for system heating and wafer hydroxylation, is the largest process energy consumer among these ALD operating activities. The break down analysis of ALD energy consumption is shown in Fig. 8.6.

As indicated, the energy consumed in the ALD process is mainly for system heating, process pumping, and experimental control. Regarding the ALD activity,

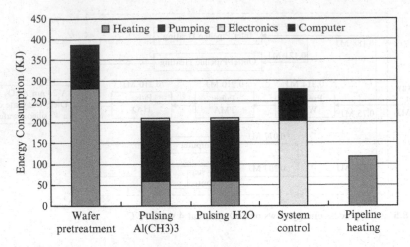

Fig. 8.6 ALD process energy consumptions

system heating in total consumes 42.7% of the energy; pumping consumes 32.8%; the electronics used for system control and data collection consume 24.5% of total energy. Regarding the 1.2 MJ energy consumed in the ALD system, 0.188 MJ is used for the system preparation and wafer pretreatment, and 0.75 MJ is consumed for operating the 200 cycle depositions, while the other 0.279 MJ is used for system monitoring and control. At a 4 s cycle time, this ALD module has a throughput of four wafers per hour for 20 nm film applications (considering 60 s of overhead time for wafer loading, system adjustment, etc.). As a result, the averaged energy consumption is approximately 0.7 MJ/wafer considering the processing capacity of such an ALD module during a typical 8 h period for batch production.

8.3.3 Material Use and Waste Generation of Nano-manufacturing

Besides energy consumption and the associated environmental impact, nano-manufacturing also consumes a lot of materials, as working materials or supplemental materials in the manufacturing processes. In particular, toxic chemicals are widely used to aid process control in nano-manufacturing. But due to the high precision requirements, among the amount of various materials put into nano-manufacturing system, only a small fraction of materials have finally turned into products. The material utilization efficiency of materials in nano-manufacturing is very low, when compared with conventional manufacturing systems. As a result, these materials input into the nano-manufacturing system but not turned into final products would be generated as process wastes, categorized as either air emission, water discharges, or land wastes.

Fig. 8.7 Human health
impact of toxic chemical
release

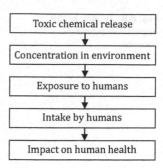

The material uses of nano-manufacturing are dependent on the characteristics and process requirements of nanotechnologies. Different nano-manufacturing technologies and processes may require different material input. In the conventional impact category, the impact of such material wastes from nano-manufacturing system could be characterized and assessed by using those conventional environmental impact assessment metrics, as presented in Chap. 3. Among these metrics, material flow analysis (MFA) is a powerful metric for understanding the material utilizations and waste generations along the nano-manufacturing process chain.

Among the different types of materials used in nano-manufacturing system, toxic chemicals are of special concerns due to their toxic effects on human health and eco-system. Toxic chemicals are heavily used in nano-manufacturing, particularly in bottom-up nano-manufacturing processes, on such applications as cleaning, etching, reacting, forming, catalyzing, etc. Toxic chemicals have been identified as a major source of environmental impact of nano-manufacturing. But in current stage of nano-manufacturing, the utilizations of toxic chemicals are inevitable. In order to green the nano-manufacturing system, the toxic chemicals should be used either at their minimum amounts or with the least impact among the candidate chemicals on the environment and eco-system.

As stated earlier in this chapter, the most effective strategy for green nano-manufacturing is to apply pollution prevention techniques to minimize its environmental impact prior to the real production. Regarding the toxic chemical control in nano-manufacturing, effective decision support tools are still needed for material screening and benchmarking during the material selection stage of nano-system design and manufacturing.

In current practice, the impact of toxic chemicals on human health is assessed based on risk analysis principles, which require consideration of the fate and transport of toxic chemical release in the environmental media, and the final exposure and impacts on public health. A simplified flow chart of toxic chemicals' impact on human health is presented in Fig. 8.7.

An innovative schematic method to characterize and benchmark the human health impact of toxic chemicals is presented here [29]. This schematic method uses a

Fig. 8.8 Human health
impact characterization
concept

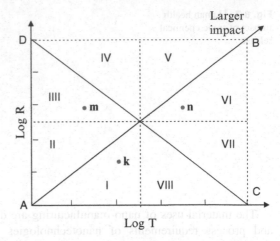

streamlined three-tiered hierarchy process which includes intake, toxicity, and persistence of a toxic chemical release for its impact characterization. In the schematic method, the human health impact of a chemical is represented by its position in a two-dimensional characterization plot, which enables the benchmarking of chemicals to be easily done by comparing the relative positions of the chemicals in the characterization plot.

In Fig. 8.8 three chemicals, m, n, k, are used as examples to demonstrate the schematic characterization of their human health impact. In the characterization plot, the two axes are both set on logarithmic scales due to the large differences of R and T magnitude. In this schematic method, the sustainable material selection of toxic chemicals can be made by benchmarking the relative positions of the candidate chemicals in this R–T two dimensional plot. The fundamental benchmarking principle is that the chemical with a higher risk and a longer persistence has a higher impact on human health. So in Fig. 8.8, those chemicals with relatively high impact on human health are characterized in the upper right corner of the characterization plot, while those with relatively low impact on human health are in the lower left corner of the plot. To facilitate rapid decision-making in material selections, the characterization plot is divided into eight regions, as shown in Fig. 8.8. Those chemicals characterized in region V and VI have higher human health impact than those in region I and II because both R and T values of chemicals in region V and VI are larger than those of chemicals in region I and II. While for those chemicals in such regions as III, IV, and VII, VIII, the tradeoffs are analyzed by evaluating the slope value of the line between the two chemicals. For two chemicals m, n, the slope value of line mn, S_{mn}, is calculated through:

$$S_{m,n} = \frac{\mathrm{Log}R_n - \mathrm{Log}R_m}{\mathrm{Log}T_n - \mathrm{Log}T_m} \qquad (8.3)$$

Table 8.1 Human health impact benchmarking of two chemicals

Condition		Result
If $R_m > R_n$ and $T_m > T_n$		$I_m > I_n$
If $R_m > R_n$ and $T_m < T_n$	$\frac{R_m}{R_n} > \frac{T_n}{T_m}$	$I_m > I_n$
	$\frac{R_m}{R_n} = \frac{T_n}{T_m}$	$I_m = I_n$
	$\frac{R_m}{R_n} < \frac{T_n}{T_m}$	$I_m < I_n$
If $R_m = R_n$ and $T_m > T_n$		$I_m > I_n$
If $R_m > R_n$ and $T_m = T_n$		$I_m > I_n$

A summary of the benchmarking principles of two chemicals' human health impact under various scenarios is presented in Table 8.1 [29].

8.3.4 Example on Conventional Waste Generations of Nano-manufacturing

This example addresses material use and waste generations from ALD nano-manufacturing processes. The example is given on the model system of ALD for deposition of Al_2O_3 high-k dielectric gate in semiconductor applications. As stated in early sections, this ALD process uses trimethyaluminum (TMA) and water as the precursor materials. Trimethyaluminum, CAS number 75-24-1, is a flammable and toxic chemical. The principal byproduct of the ALD process is methane, which is a major greenhouse gas, with a global warming potential 25 times of that of carbon dioxide.

In ALD manufacturing, the precursor materials are alternatively pulsed into the reactor and only the material deposited on the wafer surface is desired for producing economic value of the high-k dielectric gate product. The material utilization efficiency of the precursor materials, η, as defined in the following expression, is employed here for assessing the material usage of precursor materials in the ALD system:

$$\eta = \frac{\sum M_{\text{wafer}}}{\sum M_{\text{input}}} \tag{8.4}$$

The utilizations of precursor materials under three process conditions are calculated for both the trimethylaluminum and water precursors. The results are shown in Fig. 8.9 [30].

The results indicate that in this ALD operation both TMA and H_2O precursors have very low material utilization efficiencies at the three pressure levels. TMA is

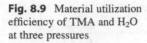

Fig. 8.9 Material utilization efficiency of TMA and H$_2$O at three pressures

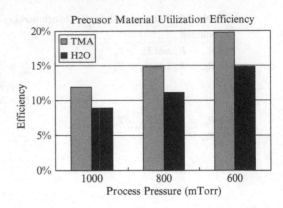

between 12 and 20%, and H$_2$O is between 8 and 15% [30]. That means more than 80% of the TMA and more than 85% of the water are wasted in current ALD operations. Increasing the material utilization efficiencies of both precursors is essential to reduce both its economic costs and environmental impacts.

As investigated, the Al$_2$O$_3$ deposition process on a silicon wafer involves a series of unstable intermediate materials at various stages of the reactions. For understanding the material flows within the ALD experimental system, the amounts of precursor materials supplied, formed and emitted from the ALD system for a 200 cycle operations at 600 mTorr are assessed and the results are shown in Fig. 8.10.

The environmental waste and emissions from an ALD system could be significant for the impact generated on the environment and human health. To understand the environmental performance of ALD technology, the amounts of TMA and water wasted from the ALD processes are shown in Figs. 8.11 and 8.12.

The wasted amounts are calculated for depositing a single Al$_2$O$_3$ film on a 4 in. silicon wafer with a typical thickness between 20 and 200 Å. Although the wasted amount seems negligible in each figure, the total wasted amount from the whole semiconductor industry would be huge, approximately at the level of 10^9 kg per year for TMA and 10^{11} l per year for deionized water, based on the 10^8 kg Al$_2$O$_3$ global annual demand of dielectric materials [31, 32].

Treatment of such process wastes from ALD operations for emission control and environmental management would be a huge burden for both the semiconductor industry and the environment. As identified, increasing the material utilization efficiency through proper ALD system design and process optimization would be viable solutions to improve both the economic and environmental performance of the ALD technology.

The principal emission from this ALD process is CH$_4$. For the amount of CH$_4$ being generated, one-third is produced from the reaction between trimethylaluminum and hydroxyls, and the other two-thirds from the reaction between H$_2$O and the absorbed trimethylaluminum, as demonstrated in Fig. 8.10. The amounts of CH$_4$ emission, based on the trimethylaluminum input, are shown in

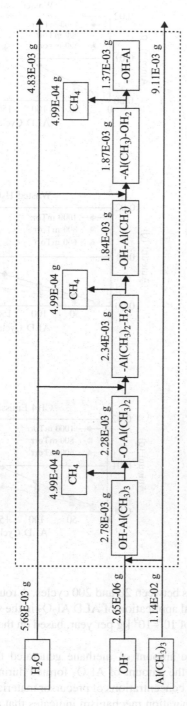

Fig. 8.10 ALD material flows of 200 cycle operations at 600 mTorr

Fig. 8.11 Wasted TMA
amount at three pressure
levels

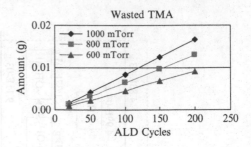

Fig. 8.12 Wasted H_2O
amount at three pressure
levels

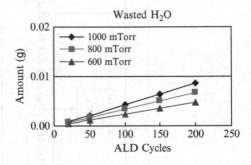

Fig. 8.13 CH_4 emissions of
ALD process at three pressure
levels

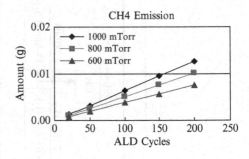

Fig. 8.13 for the ALD applications between 20 and 200 cycles. A rough estimate of the methane emissions from global application of ALD Al_2O_3 in the semiconductor industry would be in the amount of 10^8–10^9 kg per year, based on the 10^8 kg Al_2O_3 global annual demand [31, 32].

As a self-limiting process, the amount of methane generated from the ALD process is in direct proportion to the amount of Al_2O_3 formed during the reaction process, but independent of initial concentrations of precursor materials and process parameters. Such an emission generation mechanism indicates that any attempt to reduce methane emissions has to be made with a reduction of ALD operating cycles. For a targeted Al_2O_3 layer deposition, the layer thickness is fixed and the amount of methane emission is a fixed value, which can be computed through applying stoichiometric principles on the ALD deposition reactions.

8.4 Unconventional Environmental Impacts of Nano-manufacturing

The environmental impacts of nanotechnologies include not only those conventional impacts from energy and materials consumptions in the nano-manufacturing system but also a new set of impacts from nano-particle emissions. The impacts of nano-particles on the environment and eco-system are new effects from nanotechnology's application and are beyond the theoretical and instrumental limits of current environmental impact assessment practice. In order to fully characterize and assess the environmental impact of nano-manufacturing, the nano-particle emissions need to be considered together with those conventional material and energy emissions on their total environmental impact assessment. However, there are quite some challenges from both theoretical and experimental aspects of this work to include the impacts of nano-particles in the assessment.

From the nano-manufacturing system, nano-particle emissions could be produced in various processes and could be easily released into environmental media. When entering the environment, these nano-particles could be readily subject to human exposure through such exposure pathways as air, water, soil, food, etc. and through such exposure routes as inhalation, ingestion, dermal contact, etc. The impact of nano-particles would also appear on the environment and eco-system.

The impact of nano-particles on human health and eco-system could be very severe. Nano-particles and materials have been observed with more toxic effects on animals than their bulk materials due to the small particle size and large surface area to mass ratio [33]. It has also been recognized that these nanoscale particles and materials are capable to penetrate dermal barriers, cross cell membranes, breach the gas exchange regions of the lung, travel from the lung throughout the body, and interact at the molecular level [34]. It is expected that with the growing applications of nanotechnologies in industrial productions, both occupational and public exposure to the nano-particles and nano-structured materials would be dramatically increased in the near future [33].

Understanding the environmental impact of nano-particles would be very helpful in this early development stage for sustainable development of nanotechnologies. However, for the potential impact assessment of nano-particles, there are quite a few challenges which need to be scientifically solved first. At this starting stage, the impacts of nano-particles are concerned more with the health damage of human beings. For human health impact assessment of nano-particles, it is generally believed that the methods used should follow risk analysis principles. Here a few major challenges existing in the impact assessment of nano-particles on human health are listed as follows.

1. Characterization of nano-particles. The first challenge in assessing the potential impact of nano-particles is on characterization of nano-particles. Conventional environmental impact assessment is based on the mass of emissions. For the same material, a large amount of emission generates more environmental impact than a small amount. However, nano-particles' impact are not purely determined by their emission amount. Instead, the impact of nano-particles are decided by quite a number of factors including emission amount, particle size, particle number, particle shape, aspect ratio, and many others. The impact assessment of nano-particles will be based on multi-parameters, but how to scientifically characterize nano-particles for their impact assessment is still at starting stage and a lot of scientific investigations are needed in this subject.

2. Toxicity of nano-particles. A number of common nano-particles such as silver, TiO_2, carbon nanotube, etc. have been tested for their toxic effects on animals. But the toxicity of these nano-particles has not been determined due to the lack of appropriate characterization metrics and lack of scientific dose response modeling approach for nano-particle exposure. As mentioned early, the impact of nano-particles is not determined by a single parameter but by multi-parameters. In current stage, there is in great needs of developing scientific dose response modeling approach based on multi-parameters of nano-particles to determine the threshold level and safety factor of each nano-particle.

3. Exposure of nano-particles. Conventional exposures assessment is made by considering such exposure pathways as air, water, soil, food, etc. and such exposure routes as inhalation, ingestion, and dermal contact. To clarify exposure routes and pathways of nano-particles, environmental studies of nano-particles need to be conducted first to understand their fate, transport and transformations within and among various environmental medium.

4. Measurement of nano-particles. Nano-particles are so small that the measurement and characterization needs to be done with super high precision. Although the nano-particle testing techniques are under rapid development, in the current stage there is a great lack of appropriate technologies and instruments for precise and reliable measurement of nano-particles, particularly for in situ measurement in nano-manufacturing processes.

8.5 Life Cycle Assessment (LCA) of Nanotechnologies

A comprehensive environmental impact assessment of nanotechnologies needs to be conducted through LCA approaches. Among the LCA methods developed for use in the market, process-based LCA methods are preferred for nanotechnologies as they could offer detailed insight information about each specific process. But at present, those conventional LCA methods cannot be directly applied on nanotechnologies due to the needs of assessing those unconventional impacts of nanotechnologies from nano-particles. The conventional LCA methods must be adapted for application on nanotechnologies if pursing a comprehensive analysis.

As there is lack of LCA tools and methods for comprehensive assessment of environmental impact of nanotechnologies, current research conducted on LCA of nanotechnologies are mostly focused on assessing the conventional impacts, particularly on life cycle energy impact assessment. For instance, Krishnan et al. conducted a life cycle inventory analysis of nanoscale semiconductor manufacturing and found Chemical Vapor Deposition (CVD) is the largest process energy consumer (24% of total energy) and the largest greenhouse gas emission contributor [35]. Healy et al. performed a conventional environmental assessment of carbon nanotube processes and found the energy consumption (particularly the energy use in its production phase) dominates its life cycle impact [36]. Khanna et al. performed an investigation on life cycle energy consumption of carbon nano-fiber production and found that the energy requirements for carbon nano-fibers from a range of feedstock materials are 13–50 times that of primary aluminum on an equal mass basis [37].

To apply a comprehensive LCA of nanotechnologies, there are a lot of challenges which need to be resolved through rigorous research activities and investigation. By following the official framework of LCA as stated in ISO 14,000, some basic ideas are presented in the following about application and challenges in applying LCA on nanotechnologies.

8.5.1 Goal and Scope Definition in LCA of Nanotechnologies

As a first step of conventional LCA methodology, the goal and scope need to be defined. For applying LCA on nanotechnologies, the conventional practices of goal and scope definitions could still be followed. But some special characteristics of nanotechnologies should still be considered in this step before moving to the next steps for inventory and impact assessment.

8.5.2 Challenges in Inventory Analysis

Although the conventional emissions of nanotechnologies are generated along the whole LCA chain from raw material acquisition until end-of-life of the nano-based product, nano-particle emissions occur mainly in manufacturing, product use, and end-of-life stages. In order to establish a full life cycle inventory of a specific nanotechnology, those conventional emission and nano-particle emissions both need to be collected. At present, a full life cycle inventory analysis of specific nanotechnologies is unavailable. The major obstacle is the characterization of nano-particles. As stated in previous sections, conventional life cycle assessments are all conducted based on the mass of the emissions [38], but the impact of nano-particles could be linked to other factors like particle size, shape, aspect ratio, etc. A more fundamental scientific characterization of nano-particles for inventory data collection is vital for life cycle assessment of nanotechnologies.

Another challenge for life cycle assessment of nanotechnologies is due to the uncertainty of future production scales for the products using these technologies.

Most current nano-processing technologies are still under development at the laboratory scale. For a meaningful life cycle assessment, these lab scale developments need to be extrapolated to the production scale of a benchmark technology. If not, the lab scale nanotechnology will be unfairly disadvantaged by lower throughput, lower materials efficiencies, greater labor requirements, higher idle times, over-specified equipment needs, etc.

8.5.3 Challenges in Impact Analysis

Conventional life cycle impact assessment (LCIA) methods are based on life cycle impact categories. In current practice, different numbers of impact categories are employed in different methods with inherent differences using commercial tools including GaBi, Simapro, TRACI, etc. As conventional life cycle impact assessment methods are developed mainly for conventional industrial processes and production systems, the potential impact of nano-particles and the significance of the impact may not be appropriately reflected in conventional LCIA methodologies. Additional impact categories might have to be added into the category list of these various LCIA tools. This is likely to occur as research work on environmental studies of nanotechnologies shed new light on nano-particle characterization, toxicity level definition, risk of nano-particle exposure, etc. Besides, the characterization models for the new categories of nano-particle impact needs to be developed. The methods for normalization and weighing may also need to be modified.

8.6 Summary and Conclusions

Sustainable development of nanotechnology in these early stages is critical for protecting the environment and welfare of human beings in the future. By applying appropriate green manufacturing strategies in the early stages of their development, the environmental impact of nanotechnologies in future large-scale industrial production could be minimized with the least amount of environmental emissions and wastes. To enable this, environmental impact assessment methods and tools are very important in assessing and understanding the overall impact of nanotechnologies.

The environmental impact of nanotechnologies is different from those conventional technologies in that they can generate nano-particle emissions during the manufacturing processes, product use, and end-of-life. At present, it is not possible to characterize and assess the impact of nano-particles, either on the environment or on human health, due to lack of fundamental science in characterize nano-particles and lack of scientific models and methods to assess the impact of nano-particles. As a result, there are very limited studies on the environmental implications of nanotechnologies. Further, the environmental impact assessments of nanotechnologies available in public sources are focused on conventional impact of nanotechnologies, particularly on energy consumption.

There is a lot of fundamental research and science work needed in this subject in order to understand and control the environmental impact of nanotechnologies. A lot of challenges still exist particularly for applying life cycle assessment method to perform a comprehensive assessment of nanotechnologies.

References

1. National Nanotechnology Initiative (NNI) (2006) What is nanotechnology? Company website. http://www.nano.gov/html/facts/whatIsNano.html. Accessed 25 Dec 2009
2. Wetter KJ (2010) Big continent and tiny technology: nanotechnology and Africa. Foreign policy in focus, Washington, DC, October 15
3. Project on Emerging Nanotechnology (2006) Inventory of nanotechnology consumer products. http://www.nanotechproject.org/inventories/consumer/. Accessed Jan 2 2010
4. Dresselhaus MS, Dresselhaus G, Jorio A (2004) Unusual properties and structure of carbon nanotubes. Annu Rev Mater Res 34:247–278
5. Wardak A, Gorman ME, Swami N, Deshpande S (2008) Identification of risks in the life cycle of nanotechnology-based products. J Ind Ecol 12(3):1–14
6. Sweet L, Strohm B (2006) Nanotechnology-life cycle risk management. Human Ecol Risk Assess 12:528–551
7. Zhang X, Sun C, Fang N (2004) Manufacturing at nanoscale: top-down, bottom-up and system engineering. J Nanoparticle Res 6:125–130
8. Sengül H, Theis TL, Ghosh S (2008) Toward sustainable nanoproducts: an overview of nanomanufacturing methods. J Ind Ecol 12(3):329–359
9. Dowling A (2004) Nanoscience and nanotechnologies: opportunities and uncertainties. Summary and recommendations. The Royal Society & The Royal Academy of Engineering, London
10. Sequeira R, Genaidy A, Shell R, Karwowski W, Weckman G, Salem S (2006) The nano enterprise: a survey of health and safety concerns, considerations, and proposed improvement strategies to reduce potential adverse effects. Human Fact Ergonom Manuf 16(4):343–368
11. Bergeson LL, Auerbach B (2004) Reading the small print. Environ Forum 21(2):30–40
12. International Energy Agency (2008) World energy outlook, executive summary. Retrieved from http://www.worldenergyoutlook.org/2008.asp. Accessed 8 Nov 2010
13. Gutowski T, Dahmus J, Thiriez A (2006) Electrical energy requirements for manufacturing processes. 13th CIRP international conference on life cycle engineering, Leuven, Belgium May-June
14. Kushnir D, Sanden BA (2008) Energy requirements of carbon nanoparticle production. J Ind Ecol 12:360–374
15. Srituravanich W, Fang N, Sun C, Luo Q, Zhang X (2004) Plasmonic nanolithography. Nano letters 4(6):1085–1088
16. Liu Z, Steele JM, Srituravanich W, Pikus Y, Sun C, Zhang X (2005) Focusing surface plasmons with a plasmonic lens. Nano Lett 5(9):1726–1729
17. Lim B, Rahtu A, Gordon R (2003) Atomic layer deposition of transition metals. Nat Mater 2:749–754
18. Puurunen RL (2005) Surface chemistry of atomic layer deposition: a case study for the trimethylaluminum/water process. J Appl Phys 97:121301–121352
19. Sneh O, Clark-Phelps RB, Londergan AR, Winkler J, Seidel TE (2002) Thin film atomic layer deposition equipment for semiconductor processing. Thin Solid Films 402(2):248–261
20. Emmanuel AE (1993) On the definition of power factor and apparent power in unbalanced polyphase circuits with sinusoidal voltage and currents. IEEE Trans Power Deliv 8(3):841–852

21. Bottorff TE (2006) Commercial/industrial/general schedule E-19 medium general demandmetered time-of-use service. Pacific Gas and Electric Company, San Francisco, CA
22. Zhang TW, Bates SW, Dornfeld DA (2007) Operational energy use of plasmonic imaging lithography. Proceedings of the IEEE international symposium on electronics and environment, May
23. Cabrini S (Senior scientist) (2007) Center of x-ray optics and molecular foundry. Lawrence Berkeley National Laboratory, Personal communication, 19 January
24. Marrian CRK, Tennant DM (2003) Nanofabrication. J Vac Sci Technol A 21(5):5207–5215
25. Ferguson JD, Weimer AW, George SM (2000) Atomic layer deposition of ultrathin and conformal Al_2O_3. Films on BN particles. Thin Solid Films 371:95–104
26. Ott AW, Klaus JW, Johnson JM, George SM, McCarley KC, Way JD (1997) Modification of porous alumina membranes using Al_2O_3 atomic layer controlled deposition. Chem Mater 9:707–714
27. Zhang T (2007) Life cycle assessment strategies for emerging technologies: a case study in plasmonic imaging lithography. MS thesis, University of California, Berkeley
28. Matero R, Rahtu A, Ritala M, Leskela M, Sajavaara T (2000) Effect of water dose on the atomic layer deposition rate of oxide thin films. Thin Solid Films 368:1–7
29. Yuan YC, Dornfeld D (2010) Schematic method for sustainable material selection of toxic chemicals in design and manufacturing. ASME J Mech Des 132(9)
30. Yuan YC, Dornfeld D (2010) Integrated sustainability analysis of atomic layer deposition for microelectronics manufacturing. ASME J Manuf Sci Eng 132(3)
31. Sheppard L, Abraham T (2005) Evolving dielectrics in semiconductor devices. Retrieved from http://www.altfuels.com/report/SMC040B.html. Accessed 10 Sept 2009
32. Business Communications Co. (2005) Nanomaterials market by type. Retrieved from http://www.marketresearch.com/product/print/default.asp?g=1&productid=1197081. Accessed 8 Nov 2010
33. Dreher KL (2004) Health and environmental impact of nanotechnology: toxicological assessment of manufactured nanoparticles. Toxicol Sci 77:3–5
34. National Institute for Occupational Safety and Health (NIOSH) (2007) Progress toward safe nanotechnology in the workplace. Publication No. 2007–123. U.S. Department of Health and Human Services, Centers for Disease Control and Prevention, NIOSH, Washington, DC
35. Krishnan N, Boyd S, Somani A, Raoux S, Clark D, Dornfeld D (2008) A hybrid life cycle inventory of nano-scale semiconductor manufacturing. Environ Sci Technol 42(8):3069–3075
36. Healy M, Dahlben L, Isaacs J (2008) Environmental assessment of single-walled carbon nanotube processes. J Ind Ecol 12(3):376–393
37. Khanna V, Bakshi BR, Lee LJ (2008) Carbon nanofiber production: life cycle energy consumption and environmental impact. J Ind Ecol 12(3):394–410
38. ISO (International Standard Organization) (1997) ISO 14040: Environmental management—Life cycle assessment—principles and framework. International Standard ISO 14040. ISO, Geneva

Green Manufacturing Through Clean Energy Supply

9

Chris Yuan

> *The time to embrace a clean energy future is now.*
>
> Barack Obama
>
> *The stone age didn't end because we ran out of stones*
>
> Sheikh Yamani, former OPEC oil minister

Abstract

This chapter focuses on introducing the application of clean energy supply for conventional manufacturing systems to reduce energy-related environmental emissions. Three candidate energy sources, solar photovoltaic (PV), wind, and fuel cell stationary power systems are briefly introduced. The chapter then discusses the working principles and characteristics of these three clean energy technologies, the application potential of each system assessed using quantitative methods, and an example of the reduction of greenhouse gas emissions from a global automotive manufacturing system.

9.1   Introduction

Manufacturing operations are energy-intensive. All types of manufacturing processes including casting, drilling, milling, turning, grinding, etc., need energy input to actuate machines and equipment to convert raw materials into parts and products. Energy matters in green manufacturing not only because of economic costs added to the products but also because of environmental impacts associated with the energy production and supply.

C. Yuan (✉)
University of Wisconsin, Milwaukee, WI, USA
e-mail: cyuan@uwm.edu

D.A. Dornfeld (ed.), *Green Manufacturing: Fundamentals and Applications*,
DOI 10.1007/978-1-4419-6016-0_9, © Springer Science+Business Media New York 2013

The typical energy consumed in manufacturing is electricity supplied from the local electrical grid, while other forms of energy sources such as natural gas, oil, coal, etc. are also commonly used for on-site energy supply in various manufacturing operations. Electricity supplied from a local grid is the energy generated from a number of sources including coal, natural gas, nuclear, hydro, and some renewable energy technologies such as solar, wind, biomass, geothermal, etc.

As fossil fuels such as coal, natural gas, oil, etc. contain a number of chemical elements including carbon, sulfur, nitrogen, etc., burning fossil fuels in electricity production inevitably leads to generation and emissions of a number of environmental pollutants including carbon dioxide, methane, sulfur oxides, nitrogen oxides, etc. When these emissions are released into the environment, such damaging effects as global warming, acidification, eutrophication, etc., can be introduced to the environment. The level of severity of the environmental impacts depends on the amount of emissions and the chemicophysical properties of the pollutants.

As a means to encourage sustainable practices, the environmental emissions from electricity generation are allocated to the electricity end-users. In this way, the environmental emissions associated with the amount of electricity consumed in a manufacturing system are characterized as indirect emissions, while those emissions from on-site burning of fossil fuels are characterized as direct emissions.

When compared with other industrial and commercial activities, the energy consumption of manufacturing can be intensive. Depending on the manufacturing technologies and processes employed, the amount of energy input from the local grid power and on-site burning of fossil fuels can be very different. The amounts of indirect emissions and direct emissions are also different for different manufacturing systems and facilities. The amount of energy consumed in manufacturing results in significant environmental impacts from both the direct and indirect sources of emissions and has raised much concern across government, industry, and society.

In the USA, the energy consumption of the manufacturing industry is surveyed and statistically estimated for various manufacturing sectors by the US Energy Information Administration (EIA) since 1985 [1]. The EIA's manufacturing energy consumption survey was conducted initially every 3 years during 1985–1994 and afterwards was conducted and planned for every 4 years [1]. Figure 9.1 shows the energy consumption of major manufacturing sectors in 2002 and 2006, respectively, as a result of the EIA's manufacturing energy consumption survey. The data shows that there was approximately a 3.8% average energy consumption decrease in US manufacturing industry from 2002 to 2006 [1].

Depending on the geographical location, the environmental emissions resulting from energy consumption of manufacturing are quite different from region to region because of the different technologies employed in the energy supply structure. For example, China is mainly relying on coal-fired power plants for electricity generation, while nuclear power is the primary energy source of France. When compared, the energy structure of USA is a bit more diversified, with approximately half based on coal and one-fifth based on nuclear power. As a result, the differences in electricity generation technologies employed for energy supply leads to different environmental

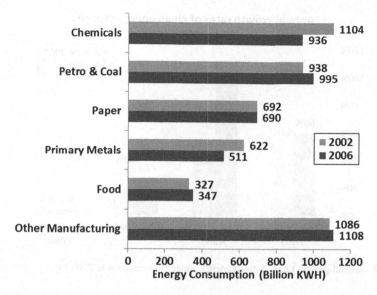

Fig. 9.1 Energy consumption of US major manufacturing sectors. *Data source*: US EIA [1]

Table 9.1 Greenhouse gas emissions of electricity generation in five major manufacturing countries from 1999 to 2002, in grams/kilowatt-hour

	USA	China	Germany	South Korea	Japan
CO_2 (g/kW h)	676	839	539	493	417
CH_4 (g/kW h)	0.01815	0.01458	0.00637	0.00758	0.00839
N_2O (g/kW h)	0.01053	0.01841	0.00779	0.00672	0.00465

Data source: US EIA [2]

emissions on the same amount of energy consumptions in a manufacturing system. This was referred to in Chap. 1 with respect to the manufacture of a typical automobile.

Take greenhouse gas emissions as an example, Table 9.1 shows the three major greenhouse gas emissions of electricity generation, per kilowatt-hour produced, in the world's five major manufacturing countries.

Reducing energy consumption of manufacturing can certainly lead to reduction of the energy-related emissions and, additionally, cut the economic costs of manufacturing systems. In the past several years, the manufacturing industry has generally taken this strategy to heart in greening their manufacturing systems through energy efficiency improvements or switching from high-polluting energy sources such as coal to low-polluting energy sources such as natural gas. However, there are natural limits to the improvement of energy efficiency in the operation of manufacturing facilities, and at this stage, it is difficult to further improve the energy efficiency of manufacturing systems to substantially reduce emissions without dramatic changes in manufacturing process technology or energy sources.

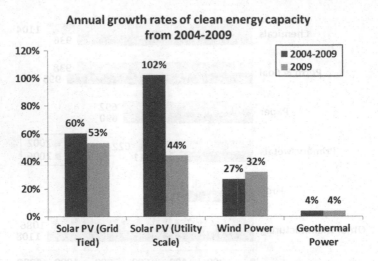

Fig. 9.2 Annual growth rates of clean energy capacity from 2004 to 2009 [3]

For reducing the energy-related emissions from manufacturing, supplying the partial energy needs of manufacturing systems through clean energy technologies has great potential since the environmental emissions of clean energy power systems are lower than those of a grid power supply, for the same amount of energy delivery. Typical clean energy technologies available on the commercial market include solar, wind, fuel cells, geothermal, and biomass.

In the past decade, these clean energy technologies have undergone rapid development and deployment internationally. Figure 9.2 shows the average annual growth rates of typical clean energy power capacity in the world during the 5-year time period between 2004 and 2009.

For large industrial scale power generation, the most popular clean energy technologies nowadays are solar photovoltaic (PV) and wind, based on the installed capacity worldwide. Statistical data shows that by the end of 2008, the installed wind power capacity has reached 121 Gigawatts (GW) worldwide, and the grid-connected solar PV capacity was 13 GW [3]. But the share of clean energy power in the global energy structure is still small when compared with fossil fuel. In 2008, 78% of global energy consumption was supplied by fossil fuel compared to only 2% from such clean power sources as solar, wind, biomass, and geothermal [3].

Application of clean energy technologies can significantly reduce the environmental emissions of manufacturing systems and, with current incentives on clean energy technologies, may also cut the economic costs a bit if appropriate clean energy power systems are selected for appropriate geographical locations [4]. For adoption of clean energy technologies in green manufacturing, there are a number of factors to be well considered in advance, especially relative to the cost benefits of the emission reduction through a clean energy supply. Since such clean energy power systems as solar, wind and geothermal are geographically dependent, selection of clean power systems also need to be considered based on the local geographical conditions.

Considering the specificity and requirements of generic manufacturing systems as well as the maturity and adaptability of clean energy technologies, such clean energy power systems as solar PV, wind and fuel cells have good potential as stationary power supplies to reduce the environmental emissions of manufacturing.

This chapter focuses on introducing the application of clean energy supply for conventional manufacturing systems to reduce the energy-related environmental emissions. The contents of this chapter are organized as follows: the working principles and characteristics of the three clean energy technologies are briefly reviewed first; then the application potential of each clean energy power system are introduced through quantitative studies. The chapter does not go into great detail on the operation of these energy systems as there are numerous other references available. The examples are shown applied to the reduction of greenhouse gas emissions and conventional air pollutants from a global automotive manufacturing system.

9.2 Clean Energy Technologies

When it comes to employing clean energy technologies to partially replace grid power supplies in manufacturing, the life cycle emissions of such clean energy technologies must be considered in the calculation of overall environmental emission reduction. In this section, the three clean energy power systems, namely solar PV, wind, and fuel cell stationary power systems are briefly introduced to aid the understanding of their application potential in green manufacturing to reduce fossil-fuel-related environmental emissions.

9.2.1 Solar Photovoltaic

Solar photovoltaic (PV) is the most popular solar technology employed in power generations. Solar PV systems use the photoelectric effect of semiconductor materials to convert sunlight directly into electricity. The major component of PV systems is the solar module, normally a number of cells connected in series. Solar PV power system produces negligible emissions during its operations and maintenance, but there are still emissions associated with other life cycle phases of a solar PV power system, including raw material acquisition, material production, manufacturing, and end-of-life.

The electricity generation of solar PV depends on the solar insulation level the PV system is exposed to, which is closely linked to the geographical conditions of the location where the PV system is deployed.

For solar PV application, the actual power output of the solar PV system under specific geographical conditions is an important measure of both the economic and environmental performance of the energy supply system. In general, the actual power output of a solar PV system in terms of AC electricity supply can be calculated through the following expression [5]:

$$AC_{out} = n_m \times I_{ave} \times A_m \times e_m \times f_{DC-AC} \qquad (9.1)$$

where

AC_{out} actual power output, AC electricity (kWh); n_m number of PV modules; I_{ave} average annual solar insolation (kWh/m^2/y); A_m surface area of one PV module (m^2); e_m module efficiency; f_{DC-AC} DC–AC conversion efficiency.

On the commercial market, there are quite a number of solar PV systems developed by various manufacturers worldwide. These solar PV modules have different technical parameters for power generations, and accordingly the PV modules selected for deployment at a specific location have direct influence on the results of the environmental impact mitigation.

9.2.2 Wind Energy

Wind is another energy source readily available in most locations. In nature, wind is another form of solar energy generated by uneven solar heating of the earth's land and sea surfaces. Wind energy resources are geographically dependent on the local conditions. Measurement of wind energy resources can be made by using wind energy density.

At a specific location, such environmental parameters as wind speed and air density (related to temperature, atmospheric pressure, and altitude) jointly determine the wind energy density. The wind speeds vary with the height above the surface of the earth, and the wind speed values obtained from the public reports are typically for 10 or 50 m height above the ground. If needed, wind speed at other heights above the ground can be obtained by using the following transformation [6]:

$$v_z = v_0 \left(\frac{z}{z_0}\right)^k \tag{9.2}$$

where

v_z Wind speed at Z m height above the ground; v_0 Wind speed at specified height of z_0; Z_0 specified height (m); k Hellman exponent.

In (9.2), the Hellman exponent value is a key parameter in converting wind speed at different heights above the ground. In general, the k value depends on the location and the shape of the terrain on the ground and the stability of the air [7]. In the city area, the $k = 0.34$ value is usually selected for the condition of neutral air above human inhibited areas [7].

As a result, the wind power density of a location can be calculated through the following expression [8]:

$$W = \frac{P}{A} = \frac{1}{2}\rho \times v_z^3 \times \Gamma\left(\frac{\lambda + 3}{\lambda}\right) \tag{9.3}$$

where

W wind power density, w/m^2; P air pressure (in units of Pascals or Newton/m^2); A area (m^2); ρ air density; v_z wind speed at z m height; λ the dimensionless Weibull shape parameter.

9.2.3   Fuel Cells

Fuel cells are electrochemical devices which consume fuel and convert the fuel energy into electricity. A fuel cell is very much similar to a battery in the way of electricity generation. The major difference between a fuel cell and a battery is that a fuel cell needs a continuous supply of fuel to produce electricity, while a battery has chemicals stored inside to react and produce electricity.

Fuel cells can be used for large-scale industrial power supply when sufficient amount of fuel can be continuously supplied to the fuel cell system. At this point, fuel cell power systems are superior to solar and wind power systems for manufacturing energy supply because the power supply is independent of local environmental conditions and can be supplied steadily for continuous operations of the manufacturing equipment and facilities.

For those stationary fuel cell power systems developed for commercial power generation, the fuel cells are usually designed to use hydrogen (or hydrogen fuels) as input to produce electricity. The application of different fuels may take the form of different chemical processes in the electricity generation. Take hydrogen as an example, the chemical reaction within the fuel cell system can be demonstrated as follows.

Anode side:

$$2H_2 \rightarrow 4H^+ + 4e^-$$

Cathode side:

$$O_2 + 4H^+ + 4e^- \rightarrow 2H_2O$$

Net reaction:

$$2H_2 + O_2 \rightarrow 2H_2O$$

But in the current stage of fuel cell application, due to the limitation of hydrogen production and storage as well as cost issues, stationary fuel cell power systems available on the commercial market are primarily built for natural gas fuel. Since natural gas is mainly composed of methane, consumption of natural gas in such fuel cell power systems leads to generation of carbon dioxide, following the reaction below:

$$CH_4 + 2H_2O = CO_2 + 4H_2.$$

If the natural gas-based fuel cell power system is to be used for energy supply for manufacturing systems, a quantitative trade-off analysis must be conducted to evaluate the net environmental emission savings of using the fuel cell power system over local grid electricity. In such a case, the life cycle emissions of the fuel cell power system must be considered in the trade-off analysis to assess the effect of environmental emission mitigations.

9.3   Application Potential of Clean Energy Supply in Green Manufacturing

The previous sections introduced the background and feasibility of using clean energy technologies to supply the power needs of manufacturing in order to reduce the environmental emissions from the use of fossil fuel-based energy. The working principles and application characteristics of solar PV, wind and fuel cells, were briefly reviewed. In this section, the application potential of these three clean energy sources is discussed for supplying the power needs of conventional manufacturing systems. The quantitative results are presented on an example of clean energy supply for global automotive manufacturing.

From previous sections in this chapter, it is clear that clean energy technologies have different application potential under different application conditions. And, accordingly the sources result in different mitigation effects on the environmental emissions reduction relative to grid power supply. In particular, such clean power systems as solar PV and wind are dependent on geographical conditions for electricity generation and need quantitative assessment of their application potential for sound decision support prior to the system deployment. For assessment of the application potential of solar and wind power systems, the geographical conditions of the installation site must be integrated in the analysis to quantify the actual power output of each clean energy supply option.

In this section, the application potential of the three clean power supply options for green manufacturing is discussed quantitatively. Assessing the application potential of clean energy supply under a specific application condition is critical for successful implementation of these systems in green manufacturing. The quantitative assessment of the application potential can be conducted in many ways. Here two quantitative methods are presented to assess clean energy application potential from two perspectives: technological performance of the supply and cost benefit of environmental emission reduction. The two aspects of assessment can be implemented together or separately depending on the actual requirement of the information for decision support in the early stages of the project planning and implementation. In particular, the cost benefit results are more valuable information and can be used directly as decision support information for strategic-planning in corporate sustainability management.

9.3.1   Technological Performance of Clean Energy Supply

Technological performance of clean energy systems is an important indicator of their application potential in power supply for manufacturing systems. But it is relatively easy to obtain the technical specifications of most of these technologies and thus convenient to assess the technological performance of clean energy supply options.

In actual application, the success of providing clean energy supply depends on a variety of factors covering the manufacturing system requirements, clean energy specifications, geographical environmental conditions, and energy sources of local grid power supply. A comprehensive assessment of all the factors for application potential of a clean energy supply typically needs multivariate decision analysis which is quite complicated and often lacks reliability in practical applications.

While the assessment of the technological performance of a clean energy supply option is important during the decision-making and strategy-planning processes, the technology assessment of clean energy power systems for a specific application can be made simply by assessing the capacity factor of the clean energy system under specific application conditions. The capacity factor of clean energy power systems is defined as the ratio between the actual power output and the total rated power of the system available for the power generations. For clean energy technologies, capacity factor can serve as a meaningful metric to assess their actual application potential in a specific geographic location for an industrial application. The capacity factor is generally defined through the following expression.

$$CF = \frac{P_{output}}{P_{system}} \times 100\% \qquad (9.4)$$

where CF capacity factor of a power supply system; P_{output} actual power output of the supply system; P_{system} rated power of the supply system.

From (9.4), we can see that the capacity factor metric actually serves as an indicator of the system efficiency for the clean energy power system when operated at different environmental conditions. For solar PV and wind power systems, the capacity factor is dependent on the technology specifications and the local environmental conditions, while the fuel cell power systems are dependent on continuous supply of fuels only. In this section, an example is shown below on the capacity factor of solar and wind power systems under different environmental conditions.

As the solar and wind power systems are dependent on the local environmental conditions for electricity generations, selection of different energy systems and installation at different sites may have a totally different energy output and emission mitigation effect. As a result, selection of sites and energy systems for installation is critical for a successful implementation of clean energy supply in green manufacturing.

For solar and wind power systems, the energy density information of installation sites and the technological specifications of the clean energy systems may serve as the basic indicators of the application potential of the clean energy power system under the local environmental condition. But more systematic assessment of the application potential requires the assessment to be conducted integrating both the local energy density information and technology performance of the clean power systems together, so as to provide accurate estimate of the system efficiency and power output of the system for sound decision support in green manufacturing applications.

Table 9.2 Representative solar PV modules for clean energy supply

Solar PV module	Rated module power (W)	Module efficiency (%)	Module surface area (m^2)
Suntech STP210-18/Ud	210	14.30	1.47
Sharp ND-224uC1	224	13.74	1.63
Qcells Q.BPPARROSOE 225	225	17.00	1.67
YingLi 210 P-26b/ 1495x990	210	14.20	1.48
Trina Solar TSM-PC05	230	14.70	1.64

Table 9.3 Representative wind turbines for clean energy supply

Wind turbine	Rated capacity (kW)	Rotor diameter (m)	Sweep area (m^2)
GE 1.5XLE	1,500	82.5	5,346.00
Sinovel 1500/77	1,500	77.4	4,705.13
Suzlon S82 1.5 MW	1,500	82	5,281.02
Nordex S77 1.5 MW	1,500	82	5,281.02

In this section, an example is shown on the calculated capacity factors of representative solar PV and wind power systems at representative locations around the world. The geographical locations selected for demonstration are representative regional manufacturing centers of global automotive manufacturers, including Detroit (USA), Mexico City (Mexico), Sao Paulo (Brazil), Shanghai (China), Cairo (Egypt), and Bochum (Germany). The technical parameters of solar PV are taken as the average of the most popular five multi-crystalline photovoltaic systems in terms of the production volume and installed capacity worldwide, as shown in Table 9.2 [5]. The technical parameters of wind power systems are taken as the average of the most popular four 1.5 MW wind turbines installed in the world, as shown in Table 9.3 [5].

Taking the average of the selected solar PV and wind power systems, the calculated capacity factors for their applications at the selected six global locations are shown in Fig. 9.3.

For the calculated results shown in Fig. 9.3, the geographical differences of solar energy among the six selected cities are characterized by the average amount of the total solar radiation incident on a horizontal surface on an annual base. The solar insolation data used in this analysis were data collected by NASA on each geographical location during a 22-year time period (from July 1983 to June 2005) [9]. The geographical differences of wind energy are characterized by the differences of wind speed and wind power density among the six selected locations. The wind speed data for the six selected cities are the data collected by NASA during a 10-year time period (from July 1983 to June 1993) [10]. The NASA wind speed data are collected only for 50 m height above the ground, while the average height of the selected four wind turbines is 80 m. In the calculation, the NASA 50 m data are transformed to 80 m wind speed data by using (9.2) as described in the previous section.

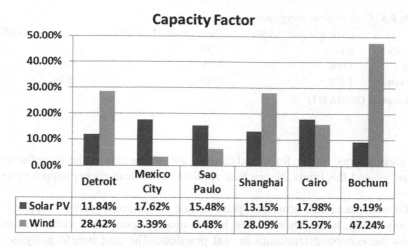

Fig. 9.3 Capacity factor of solar PV and wind power systems at the six selected global locations

The results in Fig. 9.3 clearly demonstrate that the capacity factors of such clean energy power systems as solar PV and wind, when deployed at different locations and under different environmental conditions, can be dramatically different. From the calculated capacity factor values, the application potential of such clean energy systems at a specific location can be quantitatively assessed and benchmarked. As indicated in Fig. 9.3, the quantitative results demonstrate that for current solar PV power systems, the actual capacity factors are all below 20% at these six selected locations around the world, while the capacity factor of wind can reach up to 47% among the six selected locations, depending on the local wind energy resources. Based on the quantitative results of capacity factors, the preferred installation site for solar PV and wind power systems can be easily determined from the list of potential locations. The higher the capacity factor value is, the better the application potential would be, and the better the economics of the clean energy supply.

9.3.2 Cost Benefit of Environmental Emission Mitigation Through Clean Energy Supply

The application of clean energy technologies in manufacturing for must be supported by a sound economic performance of such clean energy supply options in real practice. Since such clean energy power systems as solar PV, wind and fuel cells are produced from different technologies and processes, their economic costs are very much different. At present, the cost of electricity generation, per kilowatt-hour, through these clean energy technologies is still higher than conventional grid electricity supply, even though the costs of these clean energy technologies has decreased rapidly in the past several decades. As a result, the costs associated with

Table 9.4 Costs of clean energy power systems

	Overnight cost ($/kW)	Variable O&M ($/kW h)	Fixed O&M ($/kW)
Solar PV	6,171	0.00	11.94
Wind	1,966	0.00	30.98
Fuel cells	5,478	0.049	5.78

Data source: US EIA [11]

the clean energy supply for manufacturing are of concern to the manufacturing industry—and this limits the practical application of clean energy supply to some extent.

To understand the economic performance of potential energy supplies for manufacturing, a conventional cost benefit analysis provides a viable means to assess the economic feasibility in real practices. The cost benefit analysis can provide an estimate of the economic costs of environmental emission reductions associated with each clean energy supply option. When employed in the decision-making process, the costs of emission reduction can be used in benchmarking the economic performance of different clean energy supply options under different application conditions.

For determining the cost benefit of clean energy supplies the economic costs of electricity generations through each type of clean energy power system need to be specified first. As there are quite a number of different models available on the commercial market in each clean energy category, the costs of each clean energy technology might be slightly different with each other. In order to represent the average of the economic costs for each type of clean energy technology, here the economic costs of each clean energy power system refer to the statistically averaged economic costs data released recently by US Energy Information Administration (EIA). The costs of each clean energy system include the overnight cost and the associated variable and fixed O&M costs, as shown in Table 9.4 [11].

With the cost data as listed in Table 9.2, the economic costs of clean energy supply for manufacturing systems can be easily computed based on the scale of energy supply and the installed capacity of the clean energy power system.

For cost benefit analysis, the objective is to maximize the benefit while minimize the economic costs, if possible. The benefits of clean energy supply for manufacturing systems are the amounts of various environmental emissions which can be reduced from the local grid electricity supply which is mainly generated from fossil fuel energy around the world.

The benefits of interest here are the potential reduction in the amount of pollutants from grid electricity production compared to those from the life cycle impact of a clean energy power system. If the amount of an emission from the life cycle of a clean power system is more than that from the local grid power supply, there would be no positive benefits from the clean energy supply in this emission reduction and this option must be rejected during the decision-making process if no other benefits are to be pursued.

As a result, the cost benefit model can be constructed by considering the economic costs of each clean energy power system and the amount of each emission that can be reduced from manufacturing energy use by clean energy supply. In general, the cost benefit analysis of employing an alternate supply to reduce a specific emission from manufacturing can be conducted by using the following equation:

$$R_i = \frac{(E^i_{local} - E^i_j) \times A_j \times T_j}{(C_{Nj} + C_{Vj} \times T_j + C_{Fj})A_j} \quad (9.5)$$

where R_i the emission reduction of pollutant i, ton/\$1,000; E^i_{local} Emission factor of pollutant i from local grid power supply, kg/kWh; E^i_j the life cycle emission of pollutant i from clean energy j, kg/kWh; A_j total installed capacity of clean energy power system, j; T_j operational life time of clean power system, j, in hours; C_{Nj} overnight cost of clean power system, j, \$/kW; C_{Vj} Variable O&M cost of clean power system, j, \$/kWh; C_{Fj} Fixed O&M cost of clean power system, j, \$/kW.

Equation (9.5) can be applied generally on any emission reduction from grid electricity consumption in manufacturing by using such clean energy supplies as solar, wind, fuel cell, etc. The quantitative result, R_i, is the amount of emission reduction per unit scale of economic input or investment. It must be noted here that the results are site-dependent on geographical locations due to the difference of the emission factors of the local grid power supply and the actual power output of clean power systems.

In this section, the application of clean energy supply for conventional manufacturing systems are demonstrated on the reductions of greenhouse gas (GHG) emissions from global automotive manufacturing systems.

9.3.3 Example of Greenhouse Gas Emission Mitigation of Global Automotive Manufacturing Through Clean Energy Supply

Automotive manufacturing is very energy-intensive. Greenhouse gas emissions are generated in automotive manufacturing from both the direct on-site consumption of fossil fuel energy and indirect consumption of grid electricity. As estimated, a typical vehicle requires approximately 120 Gigajoules of energy input for its manufacture [12]. Aware of the heavy energy consumption in automotive manufacturing and the associated environmental impacts, the Alliance of Automobile Manufacturers (AAM), formed by such major global automotive manufacturers as GM, Ford, Chrysler, Toyota, Mitsubishi, Mercedes, Porsche, Volkswagen, Volvo, and Jaguar, have committed to achieve a 10% reduction in greenhouse gas emissions per number of vehicles produced from their US automotive manufacturing facilities by 2012 from a base year of 2002 [13]. During the time

Table 9.5 Selected fuel cell stationary power system for clean energy supply

Fuel cells	Manufacturer	Fuel type	Rated power (kW)
PureCell 200	UTC Power	Natural gas	200
Nedstack PS100	Nedstack	Hydrogen	100

period between 2002 and 2005, the AAM members have already reduced the GHG emissions intensity, measured as CO_2 emission per number of vehicles produced, of their US facilities by nearly 3% [13].

But further reduction of GHG emissions through conventional energy efficiency improvement and management is very difficult at this stage. Take General Motors (GM) as an example. GM facilities have participated in a total of 1,753 energy improvement projects from 1991 to 2007, which led to a total reduction of GHG emissions over 17 million metric tons CO_2 equivalent [14]. Through such efforts, GM has significantly improved the capacity utilization of its facility operations. In terms of the CO_2 emission intensity of its facilities in the USA, GM has decreased its CO_2 emission per vehicle built from 2.71 metric tons/vehicle in 1990 to 2.2 metric tons/vehicle in 2007 [14]. However, with such significant efforts in GHG mitigation, the total CO_2 emissions from GM's US facilities are still over 6.0 million metric tons each year, due to the large volume of production [14]. The recent economic downturn will have impacted this somewhat due to reduced demand but, over time, demand grows.

Clean energy supply for global automotive manufacturing provides a possible way to further reduce its greenhouse gases emissions and other fossil-fuel-energy-related emissions. Certainly there are costs associated with the benefits of environmental emission reduction through clean energy supply. In this section, a cost benefit analysis is applied to clean energy power systems of solar PV, wind and fuel cells to supply the partial energy needs of the global automotive manufacturing industry. The selected solar PV and wind power systems are shown in Tables 9.2 and 9.3. The commercial fuel cell power systems selected for study are shown in Table 9.5.

The cost benefit analyses on greenhouse gas reduction are calculated separately for solar PV, wind, natural gas fuel cells, and hydrogen fuel cells by using (9.5). The results are shown in the amount of greenhouse gas reduction per unit scale of economic input (tons GHG reduction/$1,000 economic input). The calculated results are shown in Fig. 9.4. As demonstrated by the calculated results, Fig. 9.4 shows the different GHG mitigation effects of clean energy supply patterns based on the same amount of investment scenario at the six selected regional automotive manufacturing locations around the world.

From the calculated results in Fig. 9.4, it can be seen that wind power has the greatest application potential among the three types of clean energy power systems for GHG emission mitigation, in particular at those locations with high wind energy density and/or high GHG emissions from local grid power supply. From the figure, it shows that application of wind in Bochum and Shanghai (two of the potential

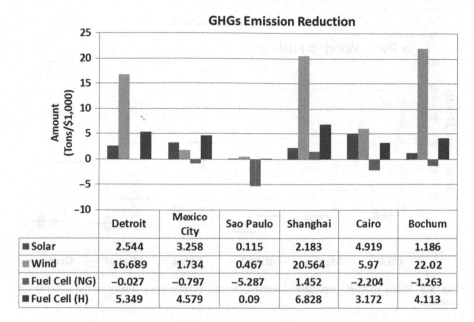

GHGs Emission Reduction						
	Detroit	Mexico City	Sao Paulo	Shanghai	Cairo	Bochum
■ Solar	2.544	3.258	0.115	2.183	4.919	1.186
■ Wind	16.689	1.734	0.467	20.564	5.97	22.02
■ Fuel Cell (NG)	−0.027	−0.797	−5.287	1.452	−2.204	−1.263
■ Fuel Cell (H)	5.349	4.579	0.09	6.828	3.172	4.113

Fig. 9.4 Economic analysis of GHG mitigation through clean energy supply

automotive manufacturing sites analyzed earlier) can achieve a GHG reduction over 20 tons per $1,000 economic input, while the application of wind in such cities as Mexico city and Sao Paulo has almost negligible mitigation effects.

From Fig. 9.4, the best GHG mitigation effects from solar PV systems are only at the level of a quarter of the mitigation effects from wind applications, based on the same scale of economic input. The economic performance of fuel cell power systems are closely related to the type of fuel used. The natural gas-based fuel cells, due to the CO_2 generation from the consumed natural gas, in most cases will further increase the total GHG emissions. The results in Fig. 9.4 indicate that GHG mitigation through natural gas-based fuel cell power system is only feasible in Shanghai, China, because of the high GHG emission factor of the local power supply [2]. In all other five locations, using natural gas-based fuel cell power system will increase the amount of GHG emissions.

The hydrogen fuel cell power system, based on the calculated results in Fig. 9.4 can achieve a mitigation effect on GHG emissions between solar and wind at the six selected locations. This is partially due to the high cost of the hydrogen-based fuel cell power system and the high energy density required for hydrogen production. It is expected that in the future hydrogen powered fuel cells would have a greater application potential as the economic costs of the systems are further lowered down and after better ways are identified for hydrogen production and storage.

The above analysis results are for applications of the selected representative clean power systems at specific geographical locations (cities). To further

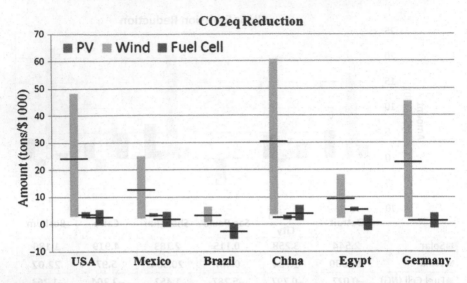

Fig. 9.5 Range of GHG mitigation potential through clean energy supply in the selected six countries

understand the GHG mitigation potential of clean energy supply for broad areas, the cost benefit analysis results for GHG mitigation through each of these clean energy supply patterns are extended to the country-wide geographical area, with the range of GHG mitigation potential assessed in the six selected countries. The range of GHG mitigation potential for solar PV and wind power systems are calculated based on the selected average technical parameters shown in Tables 9.2 and 9.3. The range of GHG mitigations from solar PV power supply is calculated by considering the best and worst power generation scenarios under the highest and lowest solar insolation conditions within the geographical boundary of the selected country. The range of GHG mitigation from wind power supply is calculated by considering the best and worst power generation scenarios under the highest wind energy density of that country and the minimum wind speed (4.47 m/s) required for wind turbine installation [15]. The range of GHG mitigation from fuel cell power supply is calculated by considering the fuel differences of the power systems based on the technical parameters shown in Table 9.5. The calculated range of GHG mitigation potential gives the maximum and minimum amount of GHGs which can be mitigated through each clean energy supply option in these six countries on the basis of the same economic input. The results are shown in Fig. 9.5, with the median value indicated for each mitigation range.

The results in Fig. 9.5 demonstrate that the best GHG mitigation opportunity is in China. With $1,000 economic investment, the maximum amount of GHG reduction can be as high as 60 tons, while application of wind power systems in the USA and Germany may also obtain a maximum of GHG reduction between 40 and 50 tons. When compared with the wind supply pattern, application of solar and fuel cell power systems has much less potential for GHG mitigation in each country. The median values of GHG mitigation range from solar and wind power supply are almost at the same level.

The successful reduction of GHG emissions through clean energy supply depends on many factors. From the results of the analysis presented here, the most important factors for an optimal GHG mitigation are the selection of the clean energy technology and the geographical location for system installation. In this analysis, the technical differences of the selected power systems in each clean energy category have not been fully assessed and benchmarked, but such differences are believed to have very small influences on the decision-making in clean energy technology selections.

Besides the greenhouse gas emission mitigation, application of clean energy supply for manufacturing can also reduce the emissions of many other environmental pollutants such as carbon monoxide, nitrogen oxides, particular matter, sulfur oxides, etc. A more detailed study of reducing the emissions of such environmental pollutants from conventional manufacturing systems through clean energy supply is presented in [4].

9.4 Summary

Manufacturing consumes a significant amount of energy and produces a number of environmental emissions from the direct and indirect use of fossil fuel energy. Green manufacturing opportunities drive the manufacturing industry to reduce the environmental impacts of their production activities as much as possible. One critical aspect of green manufacturing is reducing the energy-related environmental emissions from manufacturing (machines, processes, systems, facilities, and supply chains).

Reducing the energy consumption of manufacturing systems can cut the energy-related environmental emissions as well as bottom line energy costs. While the reduction of energy consumption through energy efficiency improvement is challenging to implement today as a number of efficiencies have been realized but, at the same time, manufacturing is still expanding around the world. As a result, alternative methods for further reduction of energy-related environmental emissions from manufacturing systems need to be explored.

This chapter reviews the possibility of supplying clean power for conventional manufacturing systems to reduce the energy-related environmental emissions. The three clean energy power systems reviewed here included solar PV, wind, and fuel cells. Their working principles and performance characteristics in supplying the energy needs of conventional manufacturing systems were reviewed.

For studying the feasibility of clean energy supply in green manufacturing, the application potential of clean energy power systems under various application conditions is used as the metric to assess the potential of clean energy supply for industrial manufacturing system. The assessment of clean energy application potential under a specific application condition can be conducted in two phases: technological and economic performance assessment.

The technological performance of clean energy is assessed by using the capacity factor of each clean energy power system at a geographical condition. The economic performance of clean energy supply for manufacturing systems is assessed by using cost benefit analysis, with a generic mathematical method presented in this chapter to assist the cost benefit analysis of each clean energy supply option.

An example is taken for clean energy supply for three selected power systems to reduce the greenhouse gases emissions of global automotive manufacturing. The application potential of each clean energy supply option is assessed by using the metrics presented in this chapter to benchmark both their technological and economic performance at different geographical conditions.

The information and results provided in this chapter can serve as decision information in the early stage of corporate sustainability management through clean energy supply, particularly in determining and benchmarking the economic costs of environmental emissions reductions using various clean energy supply. Although the example used here is for greenhouse gas emissions reduction, the methods and principles presented in this chapter can be used for other emission reductions associated with fossil fuel energy consumptions and clean energy supply for various manufacturing industries and applications.

References

1. U.S. Energy Information Administration (2007a) Manufacturing energy consumption survey. Retrieved from http://www.eia.doe.gov/emeu/mecs/contents.html. Accessed 8 Nov 2010
2. U.S. Energy Information Administration (2007b) International electricity emission factors by country, 1999–2002. Retrieved from http://www.eia.doe.gov/oiaf/1605/excel/electricity_factors_99-02country.xls. Accessed 10 Sept 2010
3. REN21 (20010) Renewables 2010 global status report. Retrieved from http://documents.rec.org/topic-areas/REN21_GSR_2010.pdf. Accessed 4 Nov 2010
4. Yuan C, Dornfeld D (2009) Reducing the environmental footprint and economic costs of automotive manufacturing through an alternative energy supply. Trans North American Manufacturing Research Institute, pp 427–434
5. Zhai Q, Cao HJ, Zhao X, Yuan C (2011) Cost benefit analysis of using clean energy supply to reduce greenhouse gas emissions of global automotive manufacturing. Energies, 4(10):1478–1494
6. Simiu E, Scanlan RH (1978) Wind effects on structures: an introduction to wind engineering, 1st edn. Wiley, New York
7. Kaltschmitt M, Streicher W, Wiese A (2007) Renewable energy: technology, economics, and environment. Springer, New York
8. Chang TJ, Wu YT, Hsu HY, Chu CR, Liao CM (2003) Assessment of wind characteristics and wind turbine characteristics in Taiwan. Renew Energy 28:851–871
9. NASA (2008) Atmospheric Science Data Center, surface meteorology and solar energy, a renewable energy resource web site. Retrieved from http://eosweb.larc.nasa.gov/cgi-bin/sse/sse.cgi?na+s01+s06#s01. Accessed 12 Oct 2010
10. NASA (2005) Atmospheric Science Data Center, surface meteorology and solar energy (Release 5 Data Set). Retrieved from http://eosweb.larc.nasa.gov/sse/global/text/10yr_wspd50m. Accessed 12 Oct 2010

11. Energy Information Administration (EIA) (2010) Cost and performance characteristics of new central station electricity generating technologies. Retrieved from http://www.eia.doe.gov/oiaf/aeo/excel/aeo2010%20tab8%202.xls. Accessed 12 Sept 2010
12. Maclean H, Lave L (1998) A life-cycle model of an automobile. Environ Policy Anal 3 (7):322A–330A
13. US Department of Energy (2007) Climate vision progress report 2007. Retrieved from http://www.doe.gov/media/Climate_Vision_Progress_Report.pdf. Accessed 2 Oct 2010
14. General Motors Company (2008) Public Policy Center Environment and Energy. Voluntary reporting of general motors corporation United States green house gas (GHG) emissions for calendar year (1990–2007). Retrieved from http://www.gm.com/corporate/responsibility/environment/reports/greenhouse_gas_emissions/ghgreport_2007.pdf. Accessed 5 Oct 2010
15. American Wind Energy Association (AWEA) (2010) Retrieved from http://www.awea.org/faq/rsdntqa.html. Accessed 13 Sept 2010

Packaging and the Supply Chain: A Look at Transportation

10

Rachel Simon and Yifen Chen

> *And Man created the plastic bag and the tin and aluminum can and the cellophane wrapper and the paper plate, and this was good because Man could then take his automobile and buy all his food in one place and He could save that which was good to eat in the refrigerator and throw away that which had no further use. And soon the earth was covered with plastic bags and aluminum cans and paper plates and disposable bottles and there was nowhere to sit down or walk, and Man shook his head and cried: 'Look at this Godawful mess.*

Art Buchwald, 1970

Abstract

In the interest of sustainability, many manufacturers have taken steps to analyze the key components of their products and processes across the span of their supply chain. Product packaging—which is an especially pervasive component, spanning across the supply chain of nearly all products—has garnered particular interest in discussions of sustainability. Packaging is not only associated with its own sourcing impacts but also influences the impacts of the product, especially in terms of the shipping impacts of the product. Several organizations have developed tools and guidelines to help manufacturers make greener packaging choices in terms of packaging. Pallet utilization is one practice for improving packaging that has been put forth in these publications and is one of the few practices that consider the impacts of packaging, not only in the context of its own supply chain but also as a component of a product. This study discusses the practice of pallet utilization and identifies the cases in which it would serve as potentially beneficial. These considerations are currently lacking in the recommendations for the adoption of pallet utilization. In addition, an overview of the current methodologies used

R. Simon (✉) • Y. Chen
1115 Etcheverry Hall, University of California, Berkeley, CA 94720, USA
e-mail: rachelrific@gmail.com

D.A. Dornfeld (ed.), *Green Manufacturing: Fundamentals and Applications*,
DOI 10.1007/978-1-4419-6016-0_10, © Springer Science+Business Media New York 2013

to evaluate the environmental impacts of transportation and to optimize the distribution of the product is provided.

10.1   Introduction

In recent years producers and retailers have shown an increased focus on sustainability. Many manufacturers, both of their own accord, and in catering to the demands of the public, have sought ways to green their products and operations. To achieve these ends, many have taken steps to analyze the key components of their products across the multiple parameters of sustainability. Product packaging has become one such component.

Packaging distinctively represents the pressures that producers face in making more sustainable sourcing choices. Of all possible product components, it is one of the most pervasive, as it spans across the supply chain of nearly all products. Packaging has its own impacts from sourcing, processing, transportation, use, and disposal, and also has the potential to influence the impacts of the product. Also, packaging suffers from a profound image problem. It serves a very significant function by protecting products during transit and storage, thereby reducing the amount of waste that would occur in its absence. This practical function goes by largely unnoticed by the consuming public. Since packaging is often the only part that remains to be disposed of after the product has been consumed, end line consumers may not recognize the purpose it served in the use phase. For them packaging is the embodiment of waste. However, the impacts of the packaging industry are not insignificant. For instance, in Europe, the packaging sector has been attributed with 3.5% of CO_2 emissions [1]. Yet, it is because the utility of packaging occurs primarily before the use phase that it is an ideal candidate for an investigation into the opportunities for sourcing and manufacturing improvements.

Packaging impacts are best analyzed through an investigation into their supply chain. These impacts consist of the direct inputs and outputs of the container, as well as their contributions to the impacts of the product. The two product life cycle phases in which packaging contributes to product impacts are the transportation and use phases. However, it is difficult to generalize the functional needs that packaging must provide over diverse products. For instance, packaging serves a very different purpose over the transport and consumption of perishable food items, as compared to the functional needs it must provide for small electronic goods. Therefore, the transportation phase of the life cycle has been identified as an ideal focus for this study.

Several organizations have developed tools and guidelines to help manufacturers make greener packaging choices. These efforts have laid a necessary foundation for packaging improvements and sustainability practices in general. Out of these works, specific trends in packaging practices have emerged. These various sustainable packaging practices have been understood and achieved with varying degrees of success.

Sustainable packaging practices can be organized into three categories, those relating to their: sourcing, production, and end-of-life. Sourcing practices refer to the type of material that goes into a package. In the context of sustainability, these types of materials include those that have been deemed as "eco-friendly," come from recycled stocks, or have been sourced from regions where there are not particular concerns about impacts to the ecosystem. Production practices are associated with how materials are processed and handled in the creation and distribution of the package. These tactics can typically only be achieved when integrated into the design of the package. Practices that fall under this category are: packaging designed to have a minimal carbon footprint; packaging that helps achieve efficiencies in distribution and logistics (e.g., pallet efficiency); multi-functional packaging that may be incorporated in with other product components; separate auxiliary refill packaging; packaging whose weight has been minimized; and packaging designed to minimize the use of specific materials—such as water or energy—in processing. End-of-life practices deal with what is done with the packaging materials after the product has been consumed. Practices that encompass the end of a package's life cycle include making it so that it can be: biodegraded, composted, recycled, or reused, specifically for another purpose. Pallet utilization and lightweighting are the only two sustainability packaging practices that address the impacts of packaging in relation to the product. The benefits of packaging lightweighting can be achieved broadly, across most products, because the practice leads to reductions simultaneously in the extraction, transport, processing, and disposal of materials, all in all resulting to reductions many times over throughout the supply chain. Alternatively, the applicability of pallet utilization is nuanced; requiring consideration of a container's weight and volume, the logistics of distribution, and the methodology used in environmental impact assessments. This study will be an exploration of these aspects and a discussion of cases in which pallet utilization can be most beneficial.

10.2 Background

10.2.1 The Packaging Supply Chain

The packaging supply chain illustrates the general complexities associated with impact assessments, as essentially two chains must be considered: (1) packaging as a contributing component to the product and (2) packaging as a standalone item that has its own sourcing and logistical aspects. The sustainability of packaging is not solely measured for its own impacts, but also for the potential it has to improve the overall sustainability of the associated product.

10.2.1.1 The Supply Chain of a Container

A typical supply chain for packaging, which includes material extraction and production, and packaging production, distribution, and end-of-life alternatives, is

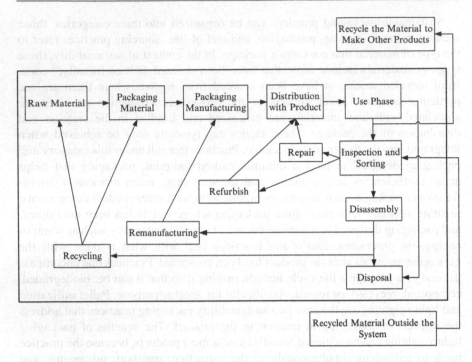

Fig. 10.1 The life cycle of packaging

depicted in Fig. 10.1. The use phase of packaging is often times considerably shorter than that of the product; where packaging's primary function is to protect the product until it is delivered to the consumer. In contrast to industrial packaging, consumer packaging is rarely used after the product has been fully consumed, although this trend is somewhat the result of consumer attitudes and behaviors. While some industrial packaging, such as stretch films, is used only once during its life cycle, most industrial packaging, such as pallets and plastic trays, are typically designed to be used numerous times in the system. These differences between industrial and consumer packaging translates into differences in the drivers and factors that affect their sustainability.

The sustainability of consumer package is mainly driven by the government regulation and societal demand. In addition, emphasis is very rarely placed on the impacts from packaging use because manufacturers have very little control on consumer behavior during this phase. The lack of recycling infrastructure is a common issue for consumer packaging. Consumers generally consider the recyclability of packaging as the main measure of sustainability [2], and as a result when firms attempt to green their packaging they tend to adopt materials that are recyclable. However, the environmental impacts of packaging cannot be evaluated based solely on its recyclability or any other single metric. Take for instance, the illustrative case of Stonyfield and their decision to switch their yogurt cup packaging material in 2001 to one that was recycled at a low rate because of the other benefits it offered. Stonyfield made this choice after conducting a life-cycle assessment (LCA), changing from high density polyethylene, which was recycled at a rate of 18.5% at the time, to polypropylene, a material with negligible recycling rates, because it was the lightest weight option for their product [3]. The switch allowed for the use of less material, as polypropylene provided containers with thinner walls that still maintained the same structural integrity, and resulted in reduced material extraction and processing, and lower transportation and waste impacts. Stonyfield compensated for the fact that most communities do not accept polypropylene for recycling, by partnering with Preserve, a producer that makes household products from 100% recycled materials, to provide consumers with locations to recycle their used yogurt cups.

In contrast to consumer packaging, industrial packaging is typically used in a closed system and kept track of through company accounting. As a result, manufacturers have a financial incentive to improve the efficiency of their industrial packaging system, which often times correlates to improved environmental impacts. The duration of the use phase for the industrial packaging life cycle can also typically be made longer than that of consumer packaging, since companies have more control of the former. Therefore, the sustainability focus for industrial packaging shifts to use efficiency and processes for recycling. In practice, many companies work with their supply chain partners on developing a more sustainable packaging system. In such collaborations manufacturers can more easily design the packaging system which reduces the overall environmental impacts. Verghese and Lewis [4] provide nine case studies in which companies work with their supply chain partners on sustainable industrial packaging system. They conclude that commercial considerations, such as the need to reduce product damage or to improve supply chain efficiency, are the main driver for changing packaging. In addition they determine that the broader benefits can usually be achieved from a well-planned and coordinated project.

An important feature of the packaging industry is that recycled materials are neither sourced from, nor made into, packaging in a closed loop system. In general, any given recycled material is collected from a variety of different products, and reprocessed in China, before it is purchased as feedstock made of recycled content [5–8]. Additionally, because the quality of packaging material degrades with each successive round of recycling, it is down cycled in most cases, meaning that it is made into products that use a lower grade of materials that are completely unlike their original form. For example, in 2008, the primary domestic end market for recycled HDPE was composite lumber, comprising 29% of demand [9]. In comparison, exports made up 57% of demand, while 4% were used in films and sheets in 2008. Trends for recycled HDPE from 2006 through 2008 can be seen in Fig. 10.2. The general trend of downcycling inhibits the possibility of recycling as a long-term solution to packaging because products cannot be infinitely made or downgraded. Eventually these systems will need the further input of raw materials.

Downcycling forms the basis for an open-loop recycling system, which makes the environmental assessment of a packaging more difficult. In LCAs many allocation procedures exist that may be used to assign the proportional shares of environmental burdens associated with a process across multiple products. However, allocation

Fig. 10.2 End markets for recovered HDPE. Adapted from Moore Recycling Associates [22–24]

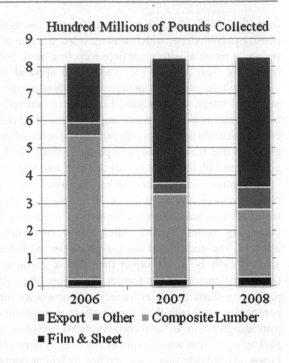

procedures are not standardized, and the lack of consensus on an optimal technique has resulted in a great degree of variability in LCA results, making the process of allocation a continued source of debate over the past two decades [10–15]. Recycling in particular has been regarded as a distinct allocation problem in need of resolution [16], as it represents both waste treatment and secondary material production. Huppes [17] presents a method to assign responsibility according to the economic value of co-products, Frischknecht [18] puts forward an allotment technique based on internal cost accounting, and Ekvall [10] suggests the use of the relative importance of supply and demand of recycled materials as market mechanisms to drive recycling as the means to assigning loads. Generally, credit is given in LCAs for avoided impacts that are associated with the displacement of virgin materials with recovered materials. Evaluations frequently show recycling to be favorable over using virgin materials, particularly for metals [19]. The allocation of recycling credits, either to the product that produces or utilizes the recycled materials, continues to be a point of contention amongst researchers [20, 21].

One simple procedure for allocation in the case of recycling has been suggested by Koltun et al. [25]. They suggested that emissions from the recycling (remanufacturing, reuse) processes and all the resources associated with it should be ascribed to the product, which is produced from these material. Figure 10.3 shows the concept behind this method. As can be seen at the end of a product's life, the environmental burdens from shipping that product to the collection and recycle facility are fully ascribed to that product. After being collected, all of the

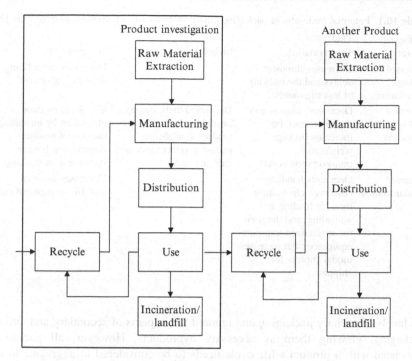

Fig. 10.3 The allocation procedure for open-loop recycling

environmental burdens from processes used to recycle and remanufacture the product into a new item are ascribed to the new product. This also means that if any recycled material is used in the initial manufacturing processes of the original product being considered, then only the environmental burdens from the recycling and remanufacturing process are counted. For any components that cannot be recycled, all the environmental burdens associated with their disposal are ascribed to the original product from which they came. This method can be easily implemented for dividing the environmental burdens of open-loop recycling. It can also avoid predicting future recycling rate and remanufacturing processes. Hence, the uncertainty of the assessments is reduced under this method.

10.2.1.2 The Container as a Component of the Product Supply Chain

In every stage of the product supply chain, packaging plays an important role in protecting it and reducing the waste from damage. Each product generally has three levels of packaging. Primary packaging holds the basic product and is what most people identify as packaging; it is sold by the retailer and disposed of by the consumer. Secondary packaging is designed to contain and protect the primary packaged product and is used to help consumers or retailers load and unload products from shelves. Tertiary packaging is designed to bundle secondary packaging units for transport between facilities. In the past, most people have focused on

Table 10.1 Potential trade-offs of packaging change with logistics activities, adapted from [26]

Packaging changes	Trade-offs		
	Transportation	Inventory	Warehousing
Increased package information	Decreases shipment delays and the tracking of lost shipments.		Decreases order filling time and labor cost
Increased package protection	Decreases damage and theft in transit, but increases package weight and transportation costs.	Decreased theft, damage, insurance; but increases product availability (sales), product value and carrying cost.	Can decrease cube utilization by increasing the size of product dimensions, but can increase it by stacking
Increased standardization	Decreases handling costs, vehicle waiting time for loading and unloading, and the need for specialized transport equipment; but increases modal choices for shippers.		Decreases material handling equipment costs

the levels of primary packaging and ignored the impacts of secondary and tertiary packaging, viewing them as necessary byproducts. However, all packaging associated with a product's life cycle needs to be considered in aggregate to get an accurate indication of the impacts of packaging. Moreover, improvements generally only occur through as a result of a trade-off between these different types of packaging. For example, the shape of the primary package may affect the numbers of product can fit into a secondary package, say, a box, and thus impacts the number of products that can be shipped in one truck. When primary packaging is reduced, secondary packaging may have to be increased if it causes the product to become more fragile. As a result the benefit of reducing primary package is someone offset by a need for more transport packaging and could cause an overall gain in the amount of packaging.

Additionally, all of the activities in a supply chain are linked and any change is accompanied by a trade-off. Gustafsson et al. [26] note that several trade-offs exist between packaging impacts and the cost of logistics, as shown in Table 10.1. While these trade-offs are focus on economic costs, they help to illustrate the nature of the relationships between the different activities. By extension, these considerations can provide useful insights for the evaluation of the sustainability of a product's packaging. For instance, if more packaging material is used to increase product protection thereby decreasing the damage rate, it will result in less total material wasted, since packaging usually requires far less material than the product. These trade-offs within the supply chain are all the more observable for secondary packaging. Take the case of a reusable/returnable transport packaging system, where the reuse rate of a tray or container depends on its durability.

Since any change to the packaging system also affect the other activities in the supply chain, the scope for the packaging assessments should incorporate these aspects of the product supply chain. Not only should resource consumption and associated impacts from packaging be included but those associated with the product also need to be recorded for further assessment. For example, if a company would like to change the packaging of their product by making it lighter in weight, the impact of these changes also need to consider the protection they provide for the product. Otherwise, even if the new packaging is made from fewer resources, the increase in product damage rates may cancel out those benefits. Packaging only comprises a small portion of the life cycle impacts of a product, any company looking to green their operations will conduct these full product assessments in order to identify opportunities. Meanwhile, companies that specialized in packaging must take the functionality of packaging into account if they are to prove the sustainability of their offerings. A possible equation for a particular packaging option is offered below:

$$\frac{\text{The life cycle environmental impact of the packaging}}{(\# \text{ of reuses}) \times (1 - \text{product damage rate}) \times (\text{packaging capacity})}$$

In LCAs, the functional unit defines the unit of the entities being examined. For example, in comparisons of plastic and paper bags, it is not useful to perform evaluations based on the unit of 1 g of bag material. Instead, more insights can be gained if units are defined by their function, say, the number of bags needed for a consumer to carry 1 kg of groceries. Lilienfeld [27] notes that one paper bag is equivalent to one and a half plastic bags. Hence, functional unit evaluations would compare the environmental impacts of 1.5 plastic bags to one paper bag. In general, we can define the functional unit for packaging applications as the amount of packaging that can protect a certain amount of product over a certain period of time. One instance would be as the amount of packaging that can protect a single product during its transportation from warehouse to retailer.

Recently, several big retailers have started to look at the environmental performance of their supply chains. For example, Wal-Mart found that 90% of the carbon emissions associated with their operations (transportation, manufacturing, farming, etc.) come from their suppliers and have started to work on greening their supply chain. As part of their sustainability initiatives, Wal-Mart has set a goal to reduce packaging by 5% of their 2008 baseline globally by 2013. Reduced packaging can not only contribute to greater environmental sustainability but can also lower transportation, inventory, and waste handling costs, and reduce the shelf and storage space needed. As part of their efforts, Wal-Mart has developed a packaging scorecard to help supplier evaluate their packaging. The scorecard will be an additional basis upon which Wal-Mart will evaluate their suppliers and as a result is pressuring suppliers to improve their packaging. While their scorecard has been subjected to criticisms, and in fact may have flaws in its methodology (see more details about metrics in Chap. 2), it serves to show that the whole supply chain is being held responsible for issues of sustainability due to the expectations of downstream customers and the regulation put forth from governments.

10.2.2 Focusing on Transportation Practices

Logistical optimization has been studied and implemented into practice for decades. It has been considered at different levels of detail, from the optimal truckload utilization, shipping frequency, and routing, to the best network design and facility locations. Across these various levels, models have been created either to minimize total cost or maximize total profit, given certain assumption on parameters at other levels. For example, the classical minimum cost network design and facility location problems usually assume a fixed unit transportation cost where the truckload utilization is given. The basic network design models take into account the trade-offs between transportation cost and facility location cost. Other constraints are included in the models, such as the maximum distance from distribution centers to retailer and the maximum number of facilities. In addition to high-level strategic network design problems, real-time routing and scheduling problems have been optimized within the field given known starting and end locations, and assumed time constraints of each shipment. For example, a minimized cost routing problem may specify a certain delivery time for each task. In some cases, companies may want to simultaneously minimize their delay and costs.

Shipment container utilization is one index of efficiency. Full truckload shipments can lower the unit transportation cost per product. However, when the demand is stable, increasing shipping quantity per shipment also implies increasing inventory holding in the facilities. Therefore, depending on the trade-off between inventory holding cost and the fixed ordering transportation cost, one may choose to order fewer products in shipments that are not full. To eliminate the aforementioned problem, companies tend to combined several products into a single order when possible. The joint shipment reduces the transportation cost but increases the complication of ordering policy and may increases the processing time.

Traditionally, cost and service are the major performance factors of logistics system. However, more recently, companies have also begun focusing on integrating sustainability as a goal. Often times the goals of sustainability coincide with the methods of lowering costs and improving service, such as in the case of better routing planning, education of ECO-driving, and coordination of shipments. In addition, several big companies have adopted changes to decrease their transportation emissions by reducing shipment frequencies, changing to lower-carbon transportation modes, and reducing the size and weight of products or its packaging. However, some of the strategies aimed at improving sustainability, may actually result in worse effects on environment. For example, changing the transportation mode may increase the lead time for shipments and further increase the energy consumption at distribution centers due to increased inventory levels. Also, reducing the size of products or packaging may not benefit the environment without ample consolidation of freight flows.

10.2.3 Sustainability Impact Assessments of Transportation

The proportional impacts of transportation vary greatly between products. In one instance, Zhang et al. [28] found that the transportation impacts contribute between 10 and 20% to the emissions for the delivery of a SolFocus concentrator PV technology to its assembly site. Pa et al. [29] found that the oceanic transport of wood pellets from British Columbia to Europe comprised about 45% of the life cycle energy consumption and contributed the greatest amount of pollutants. Meanwhile Weber and Matthews [30] note that for food items, the impacts of transportation can vary between 9 and 50%, depending on how extensive the rest of the supply chain for a particular food item is.

While the significance of transportation impacts can vary immensely between products, aggregate freight-based emissions are fairly significant at the national level. According to the EPA [31] freight transport contributes 11% of all national GHG emissions and 38% of the transportation-specific GHG emissions. In addition, the transportation sector has been a substantial contributor to increases in national emissions, accounting for 28% of the growth of all GHG emissions, and 46% of energy-related emissions, since 1990 [32].

Furthermore, the impacts of freight have become more severe over time, making the need to improve the impacts of transport all the more important. US GHG emissions from freight increased 69% between 1990 and 2005, corresponding to a 30% growth in ton-mileage during this period [33]. Bomberg et al. [33] note that most of the increase in the intensity of freight emissions can be attributed to more energy-intensive modes of transportation. The four major modes of cargo freight, namely truck, rail, air, and water, account for 60, 6, 5, and 13% of freight-related GHG emissions, corresponding to the transport of 28.5, 38.2, 0.3, and 13.0% of freight tonnage, respectively [34]. One shift that has caused a rise in emissions has been a rapid growth in air cargo, which increased 63% from 1993 to 2002, making it the fastest growing mode of freight transport despite only representing a fraction of a percent [34]. However, most of the overall growth in freight GHG intensity has been from the increase in truck market shares at the expense of other, more efficient, modes of transportation [33]. Simultaneously, there has also been a drop in the energy efficiency of freight trucks [35] due to an operational decline in efficiency [33].

The transportation impacts of a product are calculated in the inventory analysis step of LCA, as described in Chap. 3. A variety of methods are used to approximate these impacts, mostly relying on existing research on vehicle emissions factors. The GHG emissions of freight transportation can be approximated as a function of either the cargo's weight or the cargo's volume [36]. Traditionally, researchers have investigated the environmental impacts of different modes of transportation from a life cycle perspective to obtain mode-specific emission factors based on the distance and weight of shipments for specific geographic regions [37, 38]. Models have also been developed which incorporate fuel use, in addition to weight-distance data for different modes of transportation [39].

The benefit of a mode-specific weight-distance based emission factor for assessing transportation impacts is that it allows organizations a simple and standardized way to estimate the life cycle contributions from the distribution of a product. In addition the method allows for the allocation of the multiple factors that comprise a vehicle's life cycle impacts, such as fuel processing and consumption, vehicle manufacturing and maintenance, and the infrastructure needs, across the various products served by the vehicle. However, the weight-distance based methodology also has limitations. For instance the number of vehicle trips needed to transport a given number of products is not considered. To illustrate, under a weight-distance methodology, there is no differentiation between transporting 100 units in one shipment by one vehicle versus transporting them in 100 different vehicle shipments. By standardizing the baseline emissions that occur from commissioning a transportation vehicle across the miles that it travels, weight reduction is the only method that can be measured or easily identified to improve transport efficiency.

Meanwhile, a few researchers have developed methodologies to allocate the impacts of shipping according the volume the item occupies in the shipments. In studies on online shopping, the need for such an evaluation has been identified, out of the necessity to package each order individually, which causes it to take up more space in vehicles [40]. Matthews et al. [41] calculate the energy requirements to ship a book purchased through e-commerce, where transportation impacts are calculated based on volume, the distance traveled, and the fuel efficiency of the vehicle. Williams and Tagami [42] also recognize that shipping individual packages affects the energy efficiency of shipping vehicles because of the volume they take up. Their model for assessing the energy impacts for transporting each unit is similar to Matthews, but also incorporates the number of trips needed to make a successful delivery to the consumer's home. In addition, Williams and Tagami attempt to incorporate the energy impacts of vehicle production based on the proportion energy consumption associated with the production phase, as compared to the use phase of its life cycle. Also, because these models use actual fuel consumption rates, they implicitly incorporate the impacts that the freight's weight has on fuel efficiency.

In addition, extensive research has been dedicated to identifying the multiple aspects that contribute to transportation impacts. Assessments can vary dramatically based on which transportation impact factor is used, and how comprehensive researchers were in their considerations for obtaining these factors. At the base level, many researchers calculate emissions as a function of weight, distance, and fuel use based on vehicle type [30, 43–47]. However, emissions factors may vary depending on the scope of a vehicle impacts. First it should be noted that in addition to variations between vehicle types, or at the inter-modal level, there is also a great degree of variability in emissions impacts within each vehicle type or in intra-modal comparisons. Trains, planes, trucks, and ships all come in a wide range of different models, each of which have different capacities and fuel efficiencies associated with them. Further, in a mode-specific assessment of life cycle emission factors of freight transportation, Facanha and Horvath [45] incorporate the transportation infrastructure and vehicle maintenance needed to support different types of vehicles. Bomberg et al. [33] acknowledge that a vehicle's fuel economy is influenced by aerodynamic drag, rolling resistance, idling losses, accessory loads, and transmission and engine inefficiencies. However, they also identify some technologies that may mitigate some of these factors. In addition, Helms and Labrecht [48] point out that most resistance factors for ground vehicles, such as rolling resistance, gradient resistance, and acceleration resistance, are all linearly dependant on transport weight.

The assumptions about the distance of the trip also play a large role in how impacts are assessed. Foremost is the way in which the transportation distance for delivering a product is calculated. Actual miles driven, routing software, and straight line regression have all been used for these assessments. The difference between these methodologies can lead to large differences in evaluations. Impacts can also vary based on the assumption of whether a vehicle is traveling a long distance or making a short trip [48]. One factor for this, as noted by Büttner and Heyn [49] and Pearce et al. [50], is that the acceleration and deceleration of a vehicle impacts its fuel efficiency. In addition, fuel efficiency is affected by the speed at which the vehicle travels [50]. For instance, Cummins Engine Company points out that decreasing the speed of a vehicle from 65 to 60 mph can improve its fuel economy by 8% [51].

McKinnon and Edwards [52] present a framework for identifying the opportunities to reduce the environmental impacts of delivering products to retailers based on the quantity of goods. This framework, adapted from the work of McKinnon [53], maps the environmental impacts of transporting goods to a retailer based on seven factors. The first parameter is the modal split, which is the total proportion of each mode of transportation vehicle that is used to ship the freight. Another consideration is the average handling factor, or the number of times goods must be handled during their distribution, based on the structure of the logistical system and the number of channels involved. The average length of the haul, determined by the retailer's sourcing strategy as well as the efficiency of routing, also plays a role in determining the environmental impacts of freight transportation. Additionally, the aggregate amount of transport needed can be minimized through the optimal utilization of the vehicle's capacity, as determined by the average load on laden trips, the percentage of empty runs, and the vehicle's carrying capacity by weight and volume. The energy efficiency of the haul, or the energy use per distance traveled, is an important factor that has received a great deal of attention in discussions regarding the environmental impact of transportation. McKinnon and Edwards note that the energy efficiency is affected by the characteristics of the vehicle, as well as the behavior of the driver, and traffic conditions. Similarly, the pollutant content of the energy, or the amount of pollution that is created from the type of fuel that is used by a vehicle, has been identified as an important consideration in transportation decisions. Lastly, McKinnon and

Edwards note the impact of other environmental effects of the transport vehicle not associated with the consumption of energy, such as an increase in noise irritation and accidents. A representational flow diagram of this framework can be found in McKinnon and Edwards.

10.2.4 The Practice of Pallet Utilization

To address the sustainability of packaging and its potential impacts on the supply chain several organizations have developed tools and guidelines to help manufacturers make greener packaging choices. Generally, these recommendations have developed out of the best practices noted by experts and are focused on methodologies for designing or redesigning packaging. These efforts have laid a necessary foundation for further improvements in packaging and sustainability practices in general. Out of these works, general trends in sustainable packaging practices have emerged and have been understood and achieved with varying degrees of success.

Pallet utilization, also known as cube utilization, is one practice that is commonly recommended as a means for improving the sustainability of packaging. Generally, pallet utilization refers to the total space that packaged products take up and the resulting number of units that can fit onto a pallet. The practice of improving pallet utilization consists of two techniques. One method is to redesign packaging elements so that the total product shipment more efficiently utilizes the available space. Another approach is to mix products of different densities in one shipment to exploit more of the space and weight constraints for shipments simultaneously. The latter technique may also involve the redesigning of packaging elements to better achieve the objectives of improved shipping efficiency.

Shipments are restricted by both the weight and volume constraints of the vehicles they will be transport in. Very few shipments simultaneously fulfill both of these capacities, as a load either tends to reach the container's weight limitations, where the shipment is said to "weigh out," or its space restrictions, where the shipment "cubes out," but rarely does both. Although, shipments that cube out frequently only maximize the floor space of a shipping container and leave excess vertical space still unused. In most cases, manufacturers have optimized the logistics of their product shipments along the parameters of weight and size, as these efficiencies lead to cost benefits. However, the employment of these optimizations only considers the current size, shape, and weight of products and does not necessarily imply that further shipping efficiencies cannot be obtained through a redesigning of the product's packaging.

Product packaging is often not designed in a space efficient way, leaving many opportunities for improved pallet utilization to be beneficial. In a survey of the packaging for 468 products sold in selected European countries, Europen [54] found that on average about 80% of the weight of a packed pallet consisted of the product, but only around half of the volume was composed of the product. In many cases product packaging can be improved so that the total number of units that can be included in each shipment can be increased by better fitting the products into a case or by improving the composition of cases on a pallet. For example the shape of packaging can be changed to improve stacking and nesting possibilities for multiple products with more rectangular sections or a more flat surface on top. In other instances, the void space between the product and packaging can be reduced to decrease the total volume of the packaged product. These are the primary concepts behind pallet utilization.

If utilized correctly, pallet utilization can provide manufacturers with a means to achieve cost and environmental benefits. Unlike other sustainable packaging practices related to material sourcing, end-of-life options, and most production practices (with the exception of multipurpose packaging), pallet utilization is one of two sustainable packaging options that considers the influence of a container on the product. More succinctly, pallet utilization automatically affects packaging and product impacts simultaneously, unlike most sustainable packaging practices. In the case of shipments that have reached their volume capacity constraints, pallet utilization can drive down shipping impacts that have already been optimized based on the existing design. If more units are included in each shipment, it will cause reductions in the aggregate amount of fuel, labor, and vehicles needed to transport a manufactured goods.

Low density products in particular are well suited for consideration for improved pallet utilization. For instance, many snack products, such as potato chips, are packaged in bags filled with air to reduce the amount of damage that occurs during shipments. These types of products occupy the entirety of the container's space without using all of its available weight capacity. If improvements were made in the amount of space these packaged products occupy, so that more units fit into each shipment, fewer total shipments would be necessary, and the impacts associated with transit of a greater number of vehicles needed to carry shipments would be mitigated.

However, the packaging for other products, such as those that cannot be stacked, can also benefit from improved pallet utilization. In addition secondary benefits may be gained through the implementation of pallet utilization. In most cases improvements in pallet utilization will result in reductions in secondary packaging for transport and handling. For instance, as more units can be moved in each pallets and boxes, fewer of them are needed. However, there are often trade-offs between primary, secondary, and tertiary packaging, where an increase in one can lead to savings in another. Ideally, any redesign will result in an overall reduction in the total packaging needed. Additionally, the handling and storage needs for the product can also potentially be decreased. As an example, refrigerated or frozen items that are redesigned for better pallet utilization require less fuel and energy, not only for shipping but also for handling and storage as well.

However, these secondary benefits do not necessarily follow from every implementation of pallet utilization. Further, if the primary objective of an application of pallet utilization is one of these secondary benefits, packaging designers would be better served by trying to meet these primary objectives instead. For the remainder of this chapter, the focus of pallet utilization will be on the benefits that occur generally across all products and their packaging, namely: improving shipping efficiency by making better use of the available space.

Given its potential opportunities to provide benefits, pallet utilization is often recommended as a generalized way to improve the sustainability of packaging. For instance, in their packaging scorecard, Wal-Mart has assigned it 15% of the score. In addition, another 10% of their evaluations are assigned to general transportation impacts. Additional sustainable packaging guidelines also recommend pallet utilization. To address the issues of resource scarcity and transport efficiency Envirowise [55] advises improving transport efficiency by choosing packaging sizes and shapes that will maximize the case and pallet utilization. Similarly, WRAP [56] notes that packaging can be used to improve the efficiency of distribution by increasing the number of items per pallet. They also suggest changing the size or shape of the primary packaging to fit more items in a box or on a pallet. Meanwhile INCPEN's [57] *Responsible Packaging Code of Practice* suggests that packaging headspace should be kept to a minimum, but recognizes that in some cases it is needed. Later, in a joint effort with Envirowise, Incpen [58] recommended that packaging be minimized so that "the sales packs fit snugly into the transport packaging, and the transport packaging's dimensions are optimised [sic] to ensure good pallet utilization (unless weight rather than volume is the critical factor for vehicle loading)."

One popular use of pallet utilization has been a redesign of cylindrical containers which have a great amount of void space when combined in cases. For instance, in 2009, Kraft Foods redesigned their Crystal Light Drink Mix canister from a round shape to a more oval design, so that around 10% less material was used in the primary packaging and 25% less was used in the shipping trays [59], improving the pallet efficiency by 33% [60]. Bottles are particularly inefficient because even greater volumes of space around the neck of the bottles are wasted, in addition to the void space around the cylindrical body. With this in mind, the Cyprian company Cubis, in a collaborative effort with the Swedish design studio "Love for Art and Business," have developed a rectangular packaging for water bottles. Similarly, in 2008 Sam's Club and Costco stores adopted a new rectangular-shaped milk jug (see Fig. 10.4). The primary benefit of this new design was that the jugs could be stacked, thereby mitigating the need for crates to transport and store the milk. Instead, containers could be stacked layers four high on a pallet with cardboard sheets between each layer, fitting more gallons of milk on trucks and in coolers. Superior Dairy of Canton, Ohio, which initially launched the design in 1998, estimates that the new jug is 50% more space efficient than traditional containers, fitting 4.5 gallons into a cubic foot versus 3 gallons in the old design [61]. It has been estimated that these savings have reduced the number of deliveries to retailers such as Sam's club store from five to two per week [61].

In addition, several manufactures have integrated the idea of pallet utilization into their overall business strategy, giving them financial and environmental competitive advantage. Twede et al. [62] note that IKEA's success can be attributed to their practice of shipping furniture in the smallest form possible, in which it is broken down into parts for assembly by the consumer. Fully assembled furniture is

Fig. 10.4 Rectangular stackable milk jugs developed by superior dairy [65]. Reprinted with permission from Packaging World magazine and http://www.packworld.com

awkward to handle and usually cannot be done mechanically. According to Twede et al. [62], by breaking down the product into compact packs, IKEA has improved logistical efficiencies and reduced damage rates. Other companies have used pallet utilization to improve the transportation impacts of one or more products. For instance, Snyder's of Hanover, the nation's largest pretzel manufacturer, redesigned their packaging to increase the number of boxes that fit on each pallet, with the help of Georgia-Pacific's Packaging Systems Optimization (PSO) service. Snyder's also upgraded to larger shipping trailers, increasing the capacity for each of their shipments. As a result, the total greenhouse gas emissions were reduced by 32,328 lb/year [63]. Similarly, in 2007, SC Johnson reconfigured their shipments to improve logistical efficiencies. While they did not redesign their packaging, they were able to achieve savings through strategically packing multiple products on the same load to obtain the best space and weight arrangement possible for each truckload. As a result, SC Johnson was able to cut their fleet by 2,098 trucks, their GHG emissions by 1,882 tons, and their costs by $1.6 million annually [64].

While pallet utilization provides a means to improve transportation impacts, current environmental assessment methodologies may not recognize such improvements. As previously mentioned, numerous factors are considered to influence the impacts of freight transportation. While transportation impacts can be calculated along the volume of cargo, a better case can be made for the use of weight as a direct contributor to transportation-related emissions impacts. As per the current techniques commonly used to estimate transportation impacts, as described above, any potential or actual emission reductions achieved through the employment of pallet utilization may not change the outcomes of these assessments. With transportation impacts that are calculated as a function of weight, distance, and vehicle type, evaluations will often times remain unchanged despite actual reductions in shipping as the result of improved pallet utilization, unless these savings are also accompanied by changes in weight. In these cases, only manufactures that are knowledgeable about the distribution of their products and the potential benefits of pallet utilization are likely to make use of the practice to achieve improvements.

The implications of this situation are that the opportunities for pallet utilization may not be easy for manufacturers to identify, and as a result packaging designers may be misguided as to when the practice of pallet utilization would be beneficial. Pallet utilization will become an increasingly significant practice, as many manufacturers and retailers are setting goals to reduce the overall amount of their packaging and as more obvious techniques are exhausted. For this reason, any instance of packaging design for improved pallet utilization serves as examples to others in the industry at this point in time, even if the actual results are less than impressive.

However, there are many instances in which pallet utilization can actually increase transportation impacts. This is often the case when improvements in space utilization are associated with more packaging material needed for a design alternative. For instance, in the case of rectangular milk jugs, like those that have been adopted by Costco and Sam's Club, Singh et al. [66] show that it is not necessarily the practice of pallet utilization, but rather the reduction of secondary packaging, that leads to improvements in transportation impacts. Singh et al. also note that for liquids in general, weight plays a greater role than space as a factor for impacts, as trailers transporting these products "weigh out" before they "cube out." In their study of three different packaging systems—traditional jugs, jugs that have been cubed for better space utilization, and jugs that have been designed to be stackable—they found that in the case of jugs that have been cubed but were not made stackable, pallet utilization was improved but the weight transported per unit increased. This was due to the fact that each container required more material and that while the accompanying shipping crates were lighter, more of them were required, so that their total weight impacts per shipment were also higher. By switching to this container design marginal improvements of 0.8% more units per shipment could be obtained, but only at a trade-off of an increase in weight by 2.2% per unit shipped, resulting in a net increase in shipping impacts. Meanwhile the stackable containers that were designed to mitigate the need for crates resulted in 5.8% more units per shipments and an overall 5.6% weight reduction, despite the fact that the primary packaging material for the jug was increased by 43%. The details of Singh's findings can be seen in Fig. 10.5 and Table 10.2.

Given the opportunities that pallet utilization provides, it is important to distinguish between the cases where the practice is in its own right beneficial and those in which it is not pallet utilization, but the accompanying byproducts—such as weight reductions—that actually causes improvements in shipping. Otherwise, pallet utilization may be applied inappropriately, leaving all that use it vulnerable to accusations of greenwashing. The remainder of this chapter is dedicated to exploring the potential of pallet utilization and determining the specific cases in which it can be most beneficial.

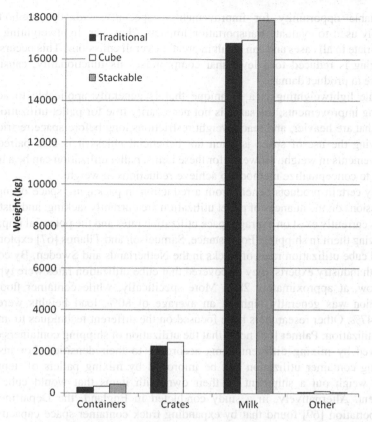

Fig. 10.5 Shipment weight aspects for different container designs for 3.79 l (1 gallon) packaging systems. Adapted from Singh et al. [66]

Table 10.2 Design differences for a 3.79 l (1 gallon) milk packaging systems

	Traditional	Cube	Cube and stackable
Total weight per shipment	19,380	19,800	18,360
Number of units per shipment	4,284	4,320	4,536
Weight per unit	4.52	4.58	4.05

Adapted from Singh et al. [66]

10.3 Recommended Method to Determine Opportunities for Improved Pallet Utilization

10.3.1 Conceptual Overview for the Applicability of Pallet Utilization

As previously noted the practice of lightweighting is generally applicable to most products; in many cases reducing the weight of packaging and products will improve their impacts. In addition, for packaging professionals lightweighting is an easily identifiable opportunity for improvements, especially with the methodology typically used to evaluate transportation impacts. However, lightweighting is not appropriate in all cases and can result in greater overall emissions. This occurs when packaging is reduced to a level that compromises its functionality, causing an increase in product damage.

While lightweighting is a technique that is generally applicable to achieve shipping improvements, the same is not necessarily true for pallet utilization. For items that are heavier, and reach weight restrictions long before space restrictions, improving the use of space is often not beneficial unless it is also paired with improvements in weight. However for these items, pallet utilization can be a helpful means to conceptualize methods to achieve reductions in weight.

Only certain products benefit from a reduction in packaging space occupancy. Discussions on the nuances of pallet utilization are currently lacking, and just a few studies currently exist on average space utilization rates and the potential impacts of improving them in shipping. For instance, Samuelson and Tilanus [67] explored the typical cube utilization rates of trucks in the Netherlands and Sweden. By consulting with industry experts, they discovered that cube utilization rates were typically very low, at approximately 28%. More specifically, while container floor area utilization was generally high, at an average of 80%, load heights were only about 47%. Other researchers have focused on the different techniques to improve cube utilization. Palmer [68] notes that the utilization of shipping containers can be improved by mixing different items according to their densities. For instance, shipping container utilization can be improved by mixing pallets of items that would weigh out a shipment on their own with items that would cube out a shipment. Alternatively, in a study conducted in England, the Department for Transportation [69] found that by expanding truck container space capacity with the use of double-decker trailers, the vehicle-kms, fuel, and CO_2 emissions for shipments were cut almost in half.

What is lacking in research and the guidelines that promote sustainability packaging practices is an identification of the cases in which pallet utilization is most appropriate. Here, we provide a conceptual methodology for designers and producers to determine these cases. The premise upon which potential packaging can be determined as ideal for improved pallet utilization is simple: focus on items whose size prevents more units from fitting on each shipment. While this concept may seem simplistic, it is not one that is specified in recommendations for the adoption of pallet utilization. Items that weigh out a shipment, rather than cube it out, will not benefit from the practice of pallet utilization, unless such applications are accompanied by secondary benefits, such as reductions in secondary or tertiary packaging. For example, it is not generally beneficial to reduce the volume of an item that weighs out a shipping container but only occupies a fraction of the available space, unless such changes are also associated with weight savings.

It should be noted that the focus of this paper is on the practice of pallet utilization in general and not the specific instances in which supplementary effects result from its employment. As was previously discussed, pallet utilization may be associated with reductions in the overall packaging weight. However, this benefit is not insured from the uptake of pallet utilization and can better be achieved by focusing on reductions in the *weight* of packaging rather than the amount of space the packaging takes up. In general, any implementation of pallet utilization that can also lead to shipping weight reductions will benefit most products, as long as the functionality of the packaging is not compromised. However this phenomenon does not occur in all cases and should not be assumed when references are made to the applicability of pallet utilization.

Below is a discussion of the applicability of pallet utilization to the different cases of single product shipment types. The instances considered are shipments that weigh out, cube out, are at floor capacity, and are partially loaded. It is assumed that very few products will cause a shipment to simultaneously max out both the space and weight constraints of a shipping container. Since these circumstances are considered to be negligible, they will not be considered in the discussion below.

10.3.2 Pallet Utilization in the Context of Shipment Density

While the applications of weight and space saving improvements are not mutually exclusive, generally one factor is more important than the other in determining the amount of products that can fit into a shipment. To discuss the concepts behind the weight and space trade-offs in shipping, consider a comparison of the density, or the ratio of weight to volume, for a product shipment and the associated shipping containers that will be used in shipments. For the product, it is assumed that the existing packaged product has already been optimized on pallets to best utilize the container capacity. For instance, if loading a pallet well below capacity allows for better stacking in the shipping container, then it is assumed that this opportunity has been exploited.

The density, or weight versus volume diagram, is shown graphically in Fig. 10.6. For any shipping containers that will be used in product transportation, the density ratio for the container's capacity is represented as the slope of the capacity line, c. Any point on line c corresponds to the percent of the respective weight and space capacity that has been occupied in a shipment. This line also intersects the point where the weight and volume capacity lines cross, represented by the top horizontal line and the far right horizontal line, respectively. Meanwhile, the density for each product can also be represented by a line, upon which the total shipment dimensions will be a point on. While the actual possible shipment weight and volume will be a set of discrete points on the line, that in general are multiplicative factors of the pallet dimensions, the density line aids in conceptualizing the relationship of the product weight and volume in comparison to this relationship in the shipping container's capacity.

Products that weigh out a shipment, such as those represented by line l_1, have a greater density slope than the container capacity line, meaning that they reach the container's weight capacity without crossing the volume capacity line. Another way to conceptualize these items is that for any given volume, the utilization rate of the weight capacity will be greater than that of the volume capacity, since the

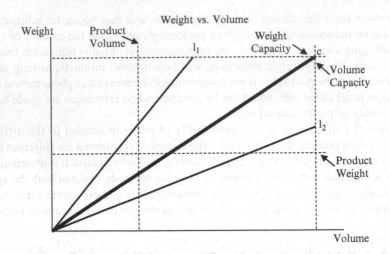

Fig. 10.6 Product and shipping container density comparisons

density ratio of weight to volume is greater than that of line c. Since weight is the limiting factor for the shipment of these items, improvements that focus on reductions in weight would be most beneficial. For these products, lightweighting, and not pallet utilization, should be the primary guiding principle for packaging improvements. Meanwhile products that cube out a shipment, such as those represented by line l_2, have a smaller slope than the capacity line. These items will cross the volume capacity line, but not the weight capacity boundary of the shipping container. Since space is the limiting factor for the shipment of these items, they are potential candidates for improved pallet utilization. It should also be noted that there is also a lower bound for any shipment, corresponding to the minimum weight and volume for shipping products without packaging. However, reducing packaging to these non-existent levels is probably not ideal, as it will most likely result in reduced marketability and increased product damage.

It should be noted that the magnitude of the density, which signifies the utilization rate of the shipping weight capacity in comparison to the utilization rate of the shipping volume capacity, does not reflect on the efficiency of the packaging. Since the density represents the relationship between the weight and the volume, any given density can simultaneously represent products with packaging that is grossly ineffi-cient as well as those with packaging that is highly optimized. In addition, shipping improvements can change the density in either direction. For example, if only the weight per unit is improved for an item that weighs out, then the density will decrease, and as a result more units will fit into each shipment, causing the volume of the shipment to increase. Meanwhile, if such an improvement in weight is also accompanied by a better utilization of the volume, the density could change in any direction, or not at all, depending on the degree of both of these improvements. Therefore, for these purposes, the density primarily signifies which aspect of the capacity is constrained, but not whether this is due to packaging efficiencies.

The density is primarily useful in identifying which constraint is at capacity and which one has excess slack, indicating which factor is more important in determining shipping rates.

10.3.3 Shipment Cases and the Applicability of Pallet Utilization

10.3.3.1 The Applications of Pallet Utilization for Shipments That Weigh Out

For products that weigh out a shipment, the potential benefits from improved pallet utilization are limited, since the greatest restrictions for these items are on their weight. In many cases, pallet utilization also results in changes to shipment weights and can actually result in a decrease in shipping efficiency. One option for items that weigh out is to use pallet utilization to mix heavier products in shipments with those that have a lower density, to better utilize the space capacity. In addition, pallet utilization can be used as a technique to conceptualize ways to lightweight the packaging for these products. In general, though, packaging improvements for items that weigh out should be focused on ways to decrease the weight, regardless of whether pallet utilization is used.

For product shipments that are at or near weight capacity, improved space efficiency can actually decrease the number of units that fit into a shipment. In many cases, improved pallet utilization will result in increases in the total shipment weight and possibly in greater weight per unit. Since pallet utilization allows for more units to fit into the same amount of space, there can also be added weight associated with these extra units. However, aggregate weight changes are also influenced by the weight of the packaging, which can vary depending upon the net change in the primary, secondary, and tertiary packaging, and may even change inversely to the total shipment weight.

To illustrate the potentially adverse effects of improved pallet utilization on a shipment whose weight has been maxed out, consider the following hypothetical scenario. Assume a truck is filled with 100 boxes, each containing 100 units, where the per unit weight of each product and all of its associated packaging is equal to eight pounds, where secondary and tertiary packaging is assigned proportionately to each product. This shipment perfectly meets the 80,000 lb weight capacity of the vehicle, but is assumed to take up only 60% of the available space. Pallet utilization can be used such that 5% more units can fit in each box. This is achieved by maintaining the original shipping case size and shape, elongating the round cylindrical form of the primary package so it better fits in the box, and adding corrugate reinforcement to each box for improved stacking strength. In total the increases in packaging are offset by the fewer number of boxes that are needed in aggregate, so that the weight per unit remains about the same at eight pounds. With this improved pallet utilization, 100 boxes, with 105 units apiece, now exceed the weight limitations of the shipping container. Instead, the maximum number of boxes that will still meet the vehicles weight constraints is 95, with the assumption that only full boxes can be included in each shipment.

Table 10.3 Example scenario illustrating the limitations of pallet utilization for weight-constrained shipments

Scenario	Number of boxes	Number of units per box	Weight per unit (lb)	Total units shipped	Total shipment weight (lb)
Original	100	100	8	1,0000	80,000
Improved pallet utilization	100	105	8	10,500	84,000
	95	105	8	9,975	79,800

With this figure, the maximum number of units that can be shipped is 9,975, which is 25 fewer than were originally in each shipment. In this case, improving pallet utilization makes shipping less efficient and results in greater transportation impacts. These figures can be seen in Table 10.3.

When there is no slack on weight constraints, any improvements in shipping quantities (i.e., an increase in the number of units per shipment) have to be proportionally offset by a reduction in weight, since the total shipment weight is equal to the product of (1) the number of boxes, (2) the number of units per box, and (3) the weight per unit. While this scenario is fictitious, it serves to illustrate how weight constraints can limit the potential benefits from pallet utilization.

As such, the focus for shipments that weigh out should be on the best methods to reduce the weight of the shipment. In many cases pallet utilization can be used as a technique to conceptualize a potential way to decrease the weight. However, weight reductions are not guaranteed with improved pallet utilization. Pallet utilization can also be used to focus on the logistics of the shipment, by combining heavy products with lighter ones that take up more space, to utilize more of this capacity. So, while possibly helpful, the potential for pallet utilization is limited for shipments that weigh out.

10.3.3.2 The Applications of Pallet Utilization for Shipments That Cube Out

For items that cube out, most of the available shipment space is occupied, and so if changes can be made to decrease the shipment's volume they are likely to be beneficial. Since the space capacity is the primary factor that determines how many units can fit into a cubed out shipment, the use of pallet utilization can improve shipping rates. If improvements in design are not a possible means to achieve these goals, pallet utilization can also be used to improve the space utilization through the mixing of products of different densities into a load. It should again be noted that, lightweighting is in general a beneficial way to improve shipping impacts, since the fuel efficiency of transport vehicles is affected by weight. However, under these circumstances, there are instances where it can actually be more advantageous to increase shipment weights. For example, if items are redesigned to take up less space, then more units could potentially fit into shipments, typically resulting in an increase in the aggregate weight of the shipment. In other instances, part of the shipment load could be replaced with different, denser, products to use more of the available space and weight capacity.

While pallet utilization is generally applicable to shipments that cube out, there exist instances in which the details of its implementation could lead to no overall improvements. Unlike products that weigh out shipments, items that cube them out have slack in their weight constraints, and as a result the options for pallet utilization can more often actually be implemented. On the other hand, similarly to shipments that weigh out, the space gained from a redesign may not actually be usable to increase the number of products that can fit into a shipment. In some instances the space gained from improvements may not be substantially enough to fit extra units. Say, for instance, that each box is decreased in height by a few inches, and then if the aggregate reduction in the height of a stack is less than that of a box, an additional row of boxes will not fit in the empty space at the top of the shipment. Depending on what changes are made, pallet utilization can decrease the total number of units that can fit into a cubed out shipment. However, this should only occur in the unlikely instances that pallet utilization is also associated with large increases in weight or when the original shipment is near the container's weight capacity. In general these trade-offs must be considered to determine the best option.

10.3.3.3 The Applications of Pallet Utilization for Shipments at Floor Capacity

Many products maximize the shipping container's available floor space, but not its available weight or space capacity. The main issue with these items is that they have structural restrictions against stacking to the full height of the container. Pallet utilization can help improve the shipping efficiency for many of these items. However, the practice alone does not provide the sole means necessary to improve stacking height for all of these types of items. Instead, deeper considerations as to why a particular item cannot be stacked must be taken into account to ascertain the best possible solution to improve shipping efficiency.

There are three main reasons why products cannot be stacked to utilize all of the available vertical space, they are either: too heavy, too fragile, or too large. Products with a high density are often too heavy to support the stacking beyond a few layers. Conversely, as products and packaging have progressively become lighter in weight, many of these items have also become too fragile to support the stacking of multiple units. Meanwhile, items that take up a lot of space may not allow for many units to simultaneously fit in the space allowed by the container. In addition it should also be noted that a producer may ship only a partial load by choice. However, the assumption for the cases being discussed here is that the best available shipping option for the existing product and packaging designs has been exploited. The applications of pallet utilization for partial load shipments will be discussed later in this chapter.

McKinnon [70] notes several reasons for declines in stacking height over time. These range from the production of lighter and smaller products that are also covered by lightweight packaging to more efficient handling equipment, and even increased health and safety regulations. For example, it has become common in sectors such as electronics, for product stacking heights to be limited by either the increased fragility of the product or the weakening of packaging materials. For these items, shipments are restricted so that they never reach the maximum volume constraint and may not reach the maximum weight constraint.

There are numerous approaches to improve the packaging of these items, so that either their stacking height can be increased or the number of units that can fit within the restricted height can be improved. Where shipping improvements can be obtained, it is likely that pallet utilization will be a helpful method, since space use is an issue for all of these items as the available space is underutilized regardless of weight. However, pallet utilization is only one of several possible techniques to increase the number of units that can fit into the available space. Adjustments can also be made to the weight of packaging so that either each unit is denser, and can potentially bear the weight of more stacked items, or that each unit is lighter, so that there is less pressure on items on the bottom of a stack. Another possible option is to reinforce packaging with additional layers, to improve the structural stability for stacking. The trade-offs between weight and space use associated with any changes must be closely monitored. No one of these particular solutions is appropriate to improve stacking in all cases, and the best solution can only be determined by a deeper consideration of the specific weight and space issues associated with any given product.

10.3.3.4 The Applications of Pallet Utilization for Partial Shipments

In addition, other logistical prerogatives, such as ordering flexibility, can take precedent over maximizing the shipping capacity. McKinnon [70] notes that order picking tends to happen early in the supply chain, causing distribution to be more demand driven. As a result pallet loads frequently consist of mixed products, making them less efficient and harder to stack. Out of this trend new challenges to remedy these shipping inefficiencies have arisen. In these cases, not only is the space and weight occupancy of an item important but also how multiple items can best be combined for a shipment. When improvements are made to partial shipments, it allows for more container capacity for other items or even the use of smaller shipping containers or vehicles. In addition, these improvements usually translate into reduced shipping costs, especially if shipping is outsourced, where the shipment's dimensions is a primary factor in pricing.

However, despite the generally positive impacts that improvements in the shipping efficiency of these items can have, the best method to achieve such efficiencies is often ambiguous. To begin with, determining how any potential changes to product and packaging dimensions will affect shipping impacts is difficult. This is due to the variability in the number of units and the type of accompanying products that will be included in any given shipment. In addition, if the improvement of the space and weight dimensions of a product allows for the inclusion of an increased number of units of another product to fit into a shipment, it is unclear how credit for these improvements is allotted between the products. What is evident for the shipment of these products is that ordering flexibility takes precedent over the impacts of shipping, and any recommendations for distribution changes should not put this option in jeopardy.

Table 10.4 The application of pallet utilization in various shipping cases

Shipment type	General approach to improve shipments
Full shipments	Focus on the capacity constraint that restricts the quantity of units that fit into a shipment. Pallet utilization will be most applicable to shipments that cube out, but can help practitioners conceptualize methods to reduce the weight of weighed out shipments.
At floor capacity shipments	The primary focus should be on the chief aspect that is preventing stacking. However, since space use is an issue, pallet utilization should be considered in all instances.
Partial shipments	Focus on whether marginal changes in weight and space can effect shipping requirements or increase the capacity available for other products. Pallet utilization can be a helpful means to conceptualize possible improvements.

For items that are commonly shipped in a partial load, it is difficult to determine the best method to achieve improved shipping impacts. If a full load of the product would cause a shipment to cube out, space use is the primary determinant of shipping capacity. For these items pallet utilization should be considered as one means to achieve greater shipping efficiency. Again, lightweighting should also be considered, as it will lead to shipping improvements for most products if they can be obtained. For products that would cause a full shipment to weigh out, the best course of action is less clear cut. The weight of these items obviously plays an important role in determining the shipping efficiency. However, the impact of these items on the total shipping efficiency of full load is unknown. The space that these items occupy could be more important than weight in determining the quantities of other products that can fit into a shipment. Therefore, in these cases, it is suggested that potential improvements in both weight and space occupancy be explored.

10.3.4 The General Applicability of Pallet Utilization

The application of pallet utilization is to redesign packaging to be more space efficient in a useful means to improve the shipping capacity for products when such options are possible. However this method is not recommended for products that weigh out shipments. Although, for these and other types of shipments, pallet utilization can be an advantageous way to pair multiple products of different densities into shipments so that they better meet the weight and space constraints of the container simultaneously. In addition, the practices of pallet utilization and lightweighting are neither mutually exclusive nor contradictory. If the shipping impacts of a product can be improved, weight-based changes will be beneficial in most situations. However, the instances in which pallet utilization will be beneficial are more limited. The instances discussed above and the potential for pallet utilization are summarized below as well as in Table 10.4.

For items that cause a shipment to weigh out, pallet utilization will be beneficial when it is applied to specifically: (1) mix products of different densities in shipments and (2) modify space utilization in a way that also results in reductions in weight. For these items, weight is a more important factor limiting the number of units per shipment, even though the amount of space that a shipment occupies can theoretically be improved. In these cases it is primarily the practice of weight reduction, not pallet utilization that will lead to improvements in shipping. Alternatively, for product shipments that space out, pallet utilization will typically provide a beneficial means to improve space efficiency, potentially even in cases it also results in increases in weight. Meanwhile, shipments at floor capacity are best improved by focusing on the particular aspects that prevent the maximum use of the available vertical space. However, with only the floor space filled, and not the volume of weight capacity constraints reached for these shipments, methods to improve the use of space—such as pallet utilization—should be considered, in addition to the specific restraints that exist on stacking. For products that will be shipped in partial loads, capacity constraints are less of an issue. Instead, marginal improvements in shipping can potentially alter shipping costs per unit, the size of shipping containers needed, and the number of units of other products that can be combined into shared shipments. For these items, the impacts of small changes in weight and space should be considered for these potential benefits.

10.4 Discussion

The complexities of the supply chain have been illustrated through a discussion of packaging. The current methodologies to assess and improve impacts have specific limitations, due, in part, to the different interpretations of sustainability. For instance, in general considerations of the functionality of components are often lacking in sustainable assessment methodologies. Also discussed here is the role that packaging plays in logistical optimization, as one of many factors influencing the transportation of freight.

Pallet utilization was identified as a practice that can provide manufacturers with a means to achieve cost and environmental benefits. Unlike most sustainable packaging practices, pallet utilization affects packaging and product impacts simultaneously. However, discussions on the nuances of pallet utilization are currently lacking, and at present require further investigation. Here a conceptual methodology for designers and producers to determine the cases in which pallet utilization is most appropriate was provided. Specifically, redesigning packaging to be more space efficient should be considered in most cases, except for products that cause shipments to weigh out. However, for these and other types of shipments, pallet utilization can be an advantageous way to pair multiple products of different densities into shipments so that they better meet the weight and space constraints of the container simultaneously.

However, while pallet utilization provides a means to improve transportation impacts, current environmental assessment methodologies may not recognize such improvements. This can lead to an improper evaluation of transportation impacts. Another implication of this situation is that the opportunities for pallet utilization may not be easy for manufacturers to identify, and as a result packaging designers may be misguided as to when the practice of pallet utilization would be beneficial.

There are also other sustainable packaging practices that may conflict with the overriding imperative of space minimization of improved pallet utilization. For instance, packaging that is made to be reusable, or out of recycled materials, often requires more material. In one study, Europen [54] found that Italian products had the lowest amount of packaging material per unit weight of product, as compared to other countries, because none of the packaging is reusable. A discussion of the comparative impacts of these conflicting practices is needed to determine which of the existing sustainable packaging practices is more advantageous.

While a substantial degree of research has been executed in the field of green and sustainable supply chains, techniques to improve them are slowly emerging. Future work will involve an assessment of the different evaluative methodologies for supply chains in general, and packaging specifically, to determine the role different considerations play in achieving more sustainable sourcing. Continued research is required on the different sustainability practices to achieve products and packaging that are truly sustainable.

References

1. Hekkert MP, Joosten LAJ, Worrell E (2000) Reduction of CO_2 emissions by improved management of material and product use: the case of transport packaging. Resour Conserv Recycl 30(1):1–27
2. Rokka J, Uusitalo L (2008) Preference for green packaging in consumer product choices—do consumers care? Int J Consum Stud 32:516–525
3. Brachfeld D, Dritz T, Kodama S, Phipps A, Steiner E, Keoleian G (2001) Life cycle assessment of the Stonyfield Farm product delivery system. Center for Sustainable Systems, University of Michigan, Ann Arbor, Michigan
4. Verghese K, Lewis H (2007) Environmental innovation in industrial packaging: a supply chain approach. Int J Prod Res 45(18):4381
5. Butler NB (2007) What's in store for plastic bags. Resour Recycl 26(6):15–19
6. Stafford B (2007) China and forest trade in the Asia-Pacific region: implications for forests and livelihoods. Environmental Aspects of China's Papermaking Fiber Supply. Prepared for Forest Trends. http://www.forest-trends.org/documents/files/doc_521.pdf. Accessed 15 Apr 2011
7. USGS (2009) Minerals yearbook 2008. US Department of the Interior, Bureau of Mines, Reston, VA
8. WRAP (2009b) The Chinese markets for recovered paper and plastics. Market situation report—Spring 2009. http://www.wrap.org.uk/downloads/China_MSR.95cc7faa.6971.pdf. Accessed 15 Apr 2011
9. American Chemistry Council (2010) The resin review. The annual statistical report for the North American plastics industry, 2010th edn. American Chemistry Council, Arlington, VA
10. Ekvall T (2000) A market-based approach to allocation at open-loop recycling. Resour Conserv Recycl 29(1–2):91–109
11. Frichknecht R (2000) Allocation in life cycle inventory analysis for joint production. Int J LCA 5:85–95
12. Reap J, Roman F, Duncan S, Bras B (2008) A survey of unresolved problems in life cycle Assessment. Part I: goals and scope and inventory analysis. Int J Life Cycle Assess 13:290–300

13. Sayagh S, Ventura A, Hoang T, François D, Jullien A (2010) Sensitivity of the LCA allocation procedure for BFS recycled into pavement structures. Resour Conserv Recycl 54(6):348–358
14. Schneider F (1994) Allocation and recycling: enlarging to the cascade system. Proceedings of the European Workshop on Allocation in LCA, Leiden, pp 39–53
15. Weidema BP (2001) Avoiding co-product allocation in life cycle assessment. J Ind Ecol 4:11–33
16. Huppes G, Schneider F (1994) Proceedings of the European workshop on Allocation in LCA. Centre of Environmental Science of Leiden University CML, Leiden, 24–25 February
17. Huppes G (1994) A general method for allocation in LCA. In: Huppes G, Schneider F (eds) Proceedings of the European workshop. Center for Environmental Sciences, Leiden University, Leiden, the Netherlands
18. Frischknecht R (1998) Life cycle inventory analysis for decision-making. scope-dependent inventory system models and context-specific joint product allocation. Ph.D. diss., Swiss Federal Institute of Technology. Zürich, ESU-services
19. Bakshi BR, Gutowski TG, Sekulic DP (2011) Thermodynamics and the destruction of resources. The Cambridge University Press, Cambridge
20. Nicholson A (2009) Methods for managing uncertainty in material selection decisions: robustness of early stage life cycle assessment. MS Tthesis, Department of Mechanical Engineering, MIT
21. Frischknecht R (2007) LCI modelling approaches applied on recycling of materials in view of environmental sustainability, risk perception and eco-efficiency. recovery of materials and energy for resource efficiency. World Congress, Davos, Switzerland, 3–5 September 2007
22. Moore Recycling Associates Inc (2008) 2006 national post-consumer recycled plastic bag and film report. Prepared for the Plastics Division of the American Chemistry Council, Arlington, VA
23. Moore Recycling Associates Inc (2009) 2007 national post-consumer recycled plastic bag and film report. Prepared for the Plastics Division of the American Chemistry Council, Arlington, VA
24. Moore Recycling Associates Inc (2010) 2008 national post-consumer recycled plastic bag and film report. Prepared for the Plastics Division of the American Chemistry Council, Arlington, VA
25. Koltun P, Tharumarajah A, Ramakrishnan S (2005) An approach to treatment of recycling in LCA study. The fourth Australian LCA Conference, Sydney, Australia
26. Gustafsson K, Jonson G, Smith DLG, Sparks, L (2006) Retailing logistics and fresh food packaging; managing change in the supply chain, London Kogan Page.
27. Lilienfeld R (2007) Review of life cycle data relating to disposable, compostable biodegradable, and reusable grocery bags. The ULS report. http://use-less-stuff.com/Paper-and-Plastic-Grocery-Bag-LCA-Summary.pdf. Accessed 1 Jul 2011
28. Zhang T, Reich-Weiser C, Nelson J, Farschi R, Minassians AD, Dornfeld D, Horne S (2008) Calculating a realistic energy payback time for a solfocus concentrator pv system. Laboratory for Manufacturing and Sustainability, UC Berkeley
29. Pa A, Craven J, Bi XT, Melin S, Sokhansanj S (2009) Evaluations of domestic applications of british columbia wood pellets based on life cycle analysis. life cycle assessment IX: toward the global life cycle economy, Boston, 29 September to 2 October
30. Weber C, Matthews HS (2008) Food-miles and the relative climate impacts of food choices in the United States. J Environ Sci Technol 42(10):3508–3513
31. Environmental Protection Agency (2006) Light-duty automotive technology and fuel economy trends: 1975 through 2006. http://www.epa.gov/otaq/fetrends.htm. Accessed 15 Apr 2011
32. EIA (2008) Annual Energy Outlook 2008 with Projections to 2030. US Department of Energy
33. Bomberg MS, Kockelman KM, Thompson MR (2008) Greenhouse gas emission control options: assessing transportation and electricity generation technologies and policies to stabilize climate change. Presented at 88th annual meeting of the transportation research board, Washington, DC

34. Frey HC, Kuo PY (2007) Best practices guidebook for greenhouse gas reductions in freight transportation. North Carolina State University. http://www4.ncsu.edu/~frey/reports/Frey_Kuo_071004.pdf. Accessed 15 Apr 2011
35. Davies J, Facanha C, Aamidor J (2007) Greenhouse gas emissions from US freight sources: using activity data to interpret trends and reduce uncertainty. Proceedings of the 87th annual meeting of the transportation research board
36. Reich-Weiser C, Dornfeld DA (2009) A discussion of greenhouse gas emission tradeoffs and water scarcity within the supply chain. J Manuf Syst 28(1):23–27
37. Facanha C (2006) Life-cycle air emissions inventory of freight transportation in the United States. Ph.D. thesis, University of California at Berkeley
38. Spielmann M, Scholz RW (2004) Life cycle inventories of transport services: background data for freight transport. Int J LCA 10(1):85–94
39. World Resources Institute (2008) GHG protocol tool for mobile combustion. Version 2.2. http://www.ghgprotocol.org/calculation-tools/all-tools. Accessed 1 Jul 2011
40. Fernie J, McKinnon AC (2009) The development of E-tail logistics. In: Fernie J, Sparks L (eds) Retail management and logistics, 3rd edn. Kogan Page, London
41. Matthews HS, Williams E, Tagami T, Hendrickson C (2002) Energy implications of online book retailing in the United States and Japan. Environ Impact Assess Rev 22:493–507
42. Williams E, Tagami T (2003) Energy use in sales and distribution via E-commerce and conventional retail. J Ind Ecol 6(2):99–114
43. "DEFRA (2005a) Guidelines for company reporting on greenhouse gas emissions annexes. Department for Environment, Food and Rural Affairs, London, July. http://archive.defra.gov.uk/environment/business/reporting/pdf/envrpgas-annexes.pdf. Accessed 4 Jan 2011
44. US Department of Energy (2004) US energy intensity indicators: transportation sector. energy efficiency and renewable energy. http://intensityindicators.pnl.gov/trend_data.stm. Accessed 15 Apr 2011
45. Facanha C, Horvath A (2007) Evaluation of life cycle air emission factors of freight transportation. J Environ Sci Technol 41:7138–7144
46. McKinnon AC (2007) CO_2 emissions from freight transport in the UK. Commission for Integrated Transport, London
47. World Business Council for Sustainable Development and World Resources Institute (2009) GHG emissions from transport or mobile sources. http://www.ghgprotocol.org/downloads/calcs/WRI_Transport_Tool.xls. Accessed 15 Apr 2011
48. Helms H, Lambrecht U (2007) The potential contribution of light-weighting to reduce transport energy consumption. Int J LCA 12(1):58–64
49. Büttner A, Heyn J (1999) Umwelt- und Kostenbilanz im Fahrzeugrohbau: Aluminium, Stahl und Edelstahl. VDI Berichte 1488. VDI, Düsseldorf, pp 137–152
50. Pearce JM, Johnson SJ, Grant GB (2007) 3D-mapping optimization of embodied energy of transportation. Resour Conserv Recycl 51:435–453
51. Cummins Engine Company (2006) Secrets of better fuel economy, the physics of MPG. http://cumminsengines.com/assets/pdf/Secrets%20of%20Better%20Fuel%20Economy_whitepaper.pdf. Accessed 15 Apr 2011
52. McKinnon AC, Edwards JB (2009) The greening of retail logistics. In: Fernie J, Sparks L (eds) Retail management and logistics, 3rd edn. Kogan Page, London
53. McKinnon AC (2008) The potential of economic incentives to reduce CO_2 emissions from goods transport. Paper prepared for the first International Transport Forum on Transport and Energy: the Challenge of Climate Change, Leipzig
54. Europen (2009) The European shopping baskets, packaging trends for fast-moving consumer goods in selected European countries. Part 1: first data collection. Brussels, Belgium
55. Envirowise (2002) Packaging design for the environment: reducing costs and quantities. GG360R. Harwel International Business Centre, Didcot, Oxfordshire
56. WRAP (2009a) A guide to evolving packaging design. www.wrap.org.uk/retail/the_guide_to_evolving_packaging_design. Accessed 15 Apr 2011

57. INCPEN (2003) Responsible packaging—code of practice for optimising packaging and minimising packaging waste. The Industry Council for Packaging and the Environment, Reading
58. Envirowise, Incpen (2008) Packguide: a guide to packaging eco-design. http://www.incpen.org/docs/PackGuide.pdf. Accessed 15 Apr 2011
59. Kraft Foods (2010) Kraft foods starts new year 150 million pounds lighter: sustainable packaging design, sourcing and partnerships remove weight equivalent of more than 150 fully loaded jumbo jets from supply chain. http://www.kraftfoodscompany.com/mediacenter/country-press-releases/us/2010/multi_media_01262010.aspx. Accessed 25 Sept 2011
60. Casey L (2009) Kraft redesigns crystal light packaging for better shelf appeal and sustainability. packaging digest. http://www.packagingdigest.com/article/367382-Kraft_redesigns_Crystal_Light_packaging_for_better_shelf_appeal_and_sustainability.php. Accessed 25 Sept 2011
61. Rosenbloom (2008) Solution, or mess? A milk jug for a green earth. The New York Times. http://www.nytimes.com/2008/06/30/business/30milk.html. Accessed 15 Apr 2011
62. Twede D, Clarke RH, Tait JA (2000) Packaging postponement: a global packaging strategy. Packag Technol Sci 13(3):105–115
63. Sterling S (2007) Transportation: driving sustainable results. Packaging World, Sustainable Outlook. http://www.packworld.com/newsletters/sp-11-14-07.html. Accessed 15 Apr 2011
64. SC Johnson & Son (2008) Doing what's right: 2008 public report. Racine, WI
65. Reynolds P (2002) Superior Likes Caseless Shipping. Packworld. http://www.packworld.com/casestudy-14588. Accessed 15 Apr 2011
66. Singh J, Krasowski A, Singh SP (2010) Life cycle inventory of HDPE bottle-based liquid milk packaging systems. Packag Technol Sci 24(1):49–60
67. Samuelson A, Tilanus B (1997) A framework efficiency model for goods transportation, with an application to regional less-than-truckload distribution. Transport Logistics 1(2):139–151
68. Palmer JM (2005) Level loading and cross docking in a global logistics network. S.M. thesis, Massachusetts Institute of Technology
69. Department for Transport (2007) Focus on double decks. Freight Best Practice Programme, HMSO, London
70. McKinnon AC (2000) Sustainable distribution: opportunities to improve vehicle loading. United Nations Environ Program Ind Environ 23(4):26–30

Enabling Technologies for Assuring Green Manufacturing

11

Athulan Vijayaraghavan and Moneer Helu

If you cannot measure it, you cannot improve it.

Lord Kelvin

Abstract

This chapter reviews various technologies applicable in characterizing the resource utilization of manufacturing processes. A review of sensors to measure and quantitatively characterize the various flows involved in manufacturing processes and machines is first presented. Given the complexity of managing and parsing the sensor data, software tools are needed to automate data monitoring and the chapter presents a framework based on event stream processing to temporally analyze the energy consumption and operational data of machine tools and other manufacturing equipment. Finally, a case study that focuses on energy measurements and demonstrates the use of energy monitoring in reasoning over the performance of a manufacturing system is presented.

11.1 Motivation

Through these past chapters, the various motivators for green manufacturing, its specific application in different manufacturing domains, and a variety of frameworks to suit particular implementations of green manufacturing have been examined. This

A. Vijayaraghavan
System Insights, 2560 Ninth Street, Suite 123A, Berkeley, CA 94710, USA
e-mail: athulan@systeminsights.com

M. Helu (✉)
Laboratory for Manufacturing and Sustainability (LMAS), Department of Mechanical Engineering, University of California, 1115 Etcheverry Hall, Berkeley, CA 94720, USA
e-mail: mhelu@berkeley.edu

D.A. Dornfeld (ed.), *Green Manufacturing: Fundamentals and Applications,*
DOI 10.1007/978-1-4419-6016-0_11, © Springer Science+Business Media New York 2013

brings the discussion to this penultimate chapter where critical technologies needed to enable green manufacturing in actual practice are identified. Like any systematic approach to manufacturing operations and process management, green manufacturing requires tools to measure and manage system performance. Critical enabling technologies for these tasks are sensors and other data collection equipment to measure resource flows and other consumption points in the manufacturing system, as well as software-based analytical techniques to apply the sensor measurements in effectively enabling and managing a green manufacturing system.

The discussion begins by looking at various sensors that can be applied in measuring resource consumption and process flows from a green manufacturing perspective.

11.2 Process Monitoring System

Critical flows of interest for manufacturing processes include electrical energy, cutting fluid, compressed air, water, and solid process waste. These flows can be divided into two categories of generalized, similarly monitored flows: electrical flows (electrical energy) and fluid flows (cutting fluid, compressed air, and water). Solid process waste flows are not included here, not because they are insignificant, but because they are normally considered as part of the standard material removal analysis. The relative importance of these flows can be determined based on the physics of the manufacturing process. For instance, many material removal-based processes occur at elevated temperatures and so will have significant fluid flows. Table 11.1 identifies critical flows for different types of manufacturing processes.

Table 11.1 Critical flows for different manufacturing systems.

Process type	Process mechanics	Critical flows
Metal cutting	Shearing (i.e., removal) of material across a sharpened edge	Electrical energy, water, oils, compressed air, solid waste, liquid waste, cutting tools
Metal forming	Reconfiguration (i.e., conservation) of material by volumetric and geometric changes to the bonding structure	Electrical energy, water, oils, solid waste, liquid waste, dies
Lithography	Formation (i.e., addition) of material layer by selective UV exposure	Electrical energy, water (DI), solid waste, liquid waste, hazardous waste, chemicals (photoresist, solidifying agents, adhesion promoters, etc.), lenses, UV light sources, masks, various gases (to fuel ovens)

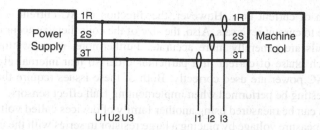

Fig. 11.1 A schematic of a 3-phase, 3-load, and 3-wire measurement used to monitor input electrical energy consumption for a machine tool

11.2.1 Electrical Flows

Monitoring electrical energy consumption can be challenging depending on the component that one wishes to monitor and whether it requires AC or DC power. All machine tools typically require 3-phase AC power from the facility source, and thus overall electrical energy consumption can be monitored through the use of a wattmeter calibrated and setup for a 3-phase, 3-load, 3-wire measurement (see Fig. 11.1). It must be emphasized that the derivation of AC power is not as straightforward as multiplying current and voltage. A description of AC power measurement is too complicated to present here, but the reader is encouraged to consult one of a number of texts on the issue such as Whitaker [1]. The wattmeter performs these calculations for the user, which allows for a straightforward measurement of electrical energy.

Once power enters the machine tool, it may be used as either AC or DC power depending on the specific component within the machine. Submetering AC power components can be accomplished using wattmeters as described previously while submetering DC power components requires only the measurement of current and voltage at the input of each component with electrical power then being the product of the measured current and voltage. The electrical energy consumed by any component is the power integrated over a time interval. It is important to mention here that submetering in practice is not as straightforward due to the complexities of a machine tool's circuitry. A challenge is that it may be difficult to know what causes a change in electrical energy consumption of any one component due to the interplay of all components within the machine tool.

Both AC and DC current can be measured using a variety of devices called ammeters that rely on the changes introduced by fluctuations in the current induced in magnetic fields. While many ammeters require that the sensor be placed into the circuit, the most effective ammeters for manufacturing applications employ the Hall effect so that they are noncontact and generally possess the highest degree of accuracy. Hall effect sensors are essentially coils placed at a known distance around a wire to be measured. These coils then output a voltage change that is induced by the change in magnetic field that is caused by a change in current in the wire. Despite their high accuracy, there are integration challenges for Hall effect sensors, particularly in AC circuits. Specifically, these sensors are unidirectional and must be placed relative to the direction of current flow. However, the direction of AC current is not always intuitive due to current reversals. Also, the size of the coils dictates accuracy: smaller diameter coils are generally more accurate. Furthermore, many wattmeters must measure each phase of current in a particular order so that internal algorithms to determine AC power are used correctly. Both of these issues require that a certain amount of testing be performed when implementing Hall effect sensors.

Voltages can be measured using another family of devices called voltmeters that essentially measure voltage by placing a large resistor in series with the circuit. This resistor allows the voltmeter to sample a small amount of current, the amount of which is proportional to the voltage drop. The accuracy of these measurements is dependent on several factors including temperature and voltage fluctuations. So, it is important to ensure constant operating conditions when performing these measurements. Also, it is recommended that Wheatstone bridge circuits be used to ensure greater precision since voltages are being measured.

11.2.2 Fluid Flows

The cumulative volume, volumetric flow rate, and mass flow rate of any fluid flow is typically monitored by a family of sensing devices called flowmeters [2]. The following discussion summarizes the use and operating principles of appropriate flowmeters for each fluid flow in machining. The reader is encouraged to consult Baker [3] for a more detailed discussion on flow measurement.

Relative to many families of sensors, flowmeters apply a wide variety of energy measurands to capture the phenomenon of interest. The most common types of flowmeters are broadly classified as flow-induced differential pressure sensors [2]. The two most popular styles of flow-induced differential pressure sensors are orifice plate meters and venturi meters. This category of flowmeters measures the difference between the pressure at point upstream, p_1, and the pressure at another point downstream, p_2, of a narrowing in the flow. If we then assume ideal incompressible volumetric flow, then the volumetric flow rate, q_v, can be estimated using the measured pressure differential, $\Delta p = p_1 - p_2$, and (11.1), which is derived from both the continuity equation and Bernoulli's equation:

$$q_v = \frac{m}{\sqrt{1 - m^2}} \left(\frac{2\Delta p}{\rho} \right)^{1/2}, \tag{11.1}$$

where $m = d^2/D^2$, D is the main pipe diameter, d is the throat diameter, and ρ is the density of the fluid.

While flow-induced differential pressure sensors have historically dominated the industrial marketplace due to the wealth of experience engineers have had with these devices as well as the general perception of their accuracy and cost, they have slowly phased out of use in many applications due to their susceptibility to unsteady and pulsating flows as well as the difficulty of their construction [2]. Other types of volume and mass flow sensors have taken the place of these flowmeters due to their higher precision and accuracy. The selection of a flowmeter from among these varied choices is highly dependent on the flow of interest. According to Baker [2], for a given set of monitoring requirements, a flowmeter should be selected based on the following properties:

- Type of fluid: Single phase (liquid or gas), slurry, or multiphase
- Special fluid constraints: Cleanliness, corrosiveness, abrasiveness, flammability, viscosity, pressure, temperature, etc.
- Environmental effects: Ambient conditions, vibrations, geometry of pipework
- Flow rate of interest: Volume or mass flow rate

Based on these considerations, three separate types of flowmeters should be used for each of the three fluid flows of interest in a machine tool or process.

11.2.2.1 Cutting Fluid

Cutting fluids are typically oil-based mixtures with a viscosity greater than water. When used in the cutting process, the volume of cutting fluid is the parameter of greatest interest since we seek to reduce chemical and resource consumption. So, positive displacement meters seem to be the most suitable type of flowmeter for cutting fluid given its physical properties.

Due to their relatively high viscosity and hydrocarbon composition, positive displacement meters are typically recommended for measuring flows of oil [2]. Positive displacement meters are volumetric flowmeters composed of gears or rotors positioned to only allow a fixed volume of fluid to pass for any one revolution of its components. The rotors can be placed in a variety of formations each with its own advantages and disadvantages depending on the fluid of interest. Example rotor configurations for positive displacement type flowmeters includes multi-rotor, oval gear, nutating disc, sliding vane, and helical rotor.

Because fluid is carried through what are essentially self-contained compartments in the flowmeter, the cumulative volume of fluid passed through is directly proportional to the number of rotations of the rotors [2]. The flow rate of the fluid can also be determined because it is directly proportional to the rotational speed of the rotors. While accuracy is dependent on the specific rotor configuration, both the cumulative volume of fluid and the fluid flow rate can be determined with typically high accuracy for positive displacement devices. These flowmeters are so well known for their accuracy that these are the sensors of choice for the metering of valuable fluids such as fuel, oil, milk, and water. In fact, generally speaking, only magnetic flowmeters are capable of greater accuracy, but magnetic devices can only monitor conducting liquids.

Accuracy is not the only benefit of positive displacement devices, nor is it a major reason why these are typically used to measure and control oil hydraulics. Other advantages of positive displacement devices include its relatively large measurement range, its lack of sensitivity to upstream flow profiles, and its ability to easily relate pulse outputs (typical of flowmeters) to the flow rate [2]. Furthermore, of the main types of flowmeters, positive displacement devices best handle higher viscosity liquids.

Positive displacement devices do have certain disadvantages as well including a relatively bulky structure, an ability to create flow pulsation downstream of the sensor or a flow blockage at the sensor, and sensitivity to sudden flow changes (especially in terms of pressure) [2]. Many of these concerns, though, are either already indirectly addressed by typical machine tool design and use (e.g., cutting fluid is already filtered to remove any particulates that could clog or damage the lines or sensor) or inconsequential to machine tool use (e.g., bulky structures are fine as long as they do not interfere with the process and flow pulsation will only result in problems if a part does not see enough cutting fluid). A specific issue that would need to be addressed, though, is that control valves should be used both downstream and upstream so that the acceleration and deceleration of cutting fluid is low enough to not cause excessive wear on the rotors and/or throw off the calibration. Furthermore, it may be necessary to simultaneously monitor the temperature of the flow to remove any thermal effects on the density of the cutting fluid and normalize the measurements made by the sensor.

11.2.2.2 Compressed Air

Similarly to cutting fluid, the consumption of compressed air is of greatest interest for cutting processes. While many flowmeters can measure gaseous flows, mass flowmeters are specifically suited for measuring the mass of any fluid and work particularly well with gases. Mass flowmeters based on convective heat transfer (or thermal mass flowmeters) are well suited to monitor compressed air flows as they are simple to implement and readily available on the commercial market. These types of mass flowmeters generally employ Equation (11.2) to determine the mass flowrate, q_m, from the measured temperature rise in the fluid flow, ΔT, due to an applied heat transfer, Q_h:

$$q_m = \frac{Q_h}{K c_p \Delta T},$$

(11.2)

where K is a constant determined by calibration, and c_p is the specific heat of the fluid at a constant pressure [2].

There are two main types of thermal mass flowmeters for gaseous fluids: capillary thermal mass flowmeter (CTMF) and insertion/in-line thermal mass flowmeter (ITMF) [2]. The general operating principle of both types of mass flowmeters is that the mass flowrate is proportional to amount of heat dissipated by the flow. To measure the heat dissipation, CTMF devices heat a central location of the flow and take measurements using thermocouples before and after the heat transfer location at points equidistant from this location. Equation (11.2) can then be used to determine the mass flowrate from the temperature change. Alternatively, ITMF devices insert two thermal sensors into the gaseous flow: one that measures the temperature of the flow for reference and the other that is kept at 20°C above the flow temperature. The mass flowrate is then proportional to the amount of heat required to keep the second sensor at 20°C above the flow temperature. For ITMF devices, Equation (11.2) must be modified as follows:

$$Q_h = k(1 + Kq_m^n)\Delta T, \tag{11.3}$$

where k is a constant that allows for heat transfer and temperature difference at zero flow, K is another constant that incorporates the area of the duct, and n is power constant related to the flow that varies between ½ and ⅓.

Both CTMF and ITMF devices have the advantage of being capable of measuring a wide range of mass flowrates, which differs from other flowmeters that tend to have difficulty measuring especially low flowrates [2]. Of the two, CTMF devices offer higher accuracy while ITMF devices tend to be much cheaper. While the difference in accuracy may not seem especially substantial, ITMF devices may be more influenced by the nature of the fluid being measured. Really, though, both are likely affected by the heat transfer rate, which depends on the fluid as well as pipe condition.

When installing either type of flowmeter, there are certain precautions that must be observed. Both devices must be installed such that the heat transfer and thermal measurement sites are insulated from the rest of the pipe to ensure that the change in measured temperature is due only to the amount of heat supplied [2]. Also, calibration should be performed as close to operating conditions as possible to remove any other sources of thermal fluctuations. For the CTMF device in particular, if a bypass arrangement is used (needed when the main flow is too large, which requires that a sample of it should be directed to an adjacent capillary tube where the CTMF is installed), then flow must be laminar to ensure that the measurement remains independent of flow profile. Lastly, CTMF devices typically require "clean" (i.e. no particulate matter), single-phase flows while ITMF devices may measure "dirty," single-phase flows. Since compressed air used in machine tools is expected to be particulate free, though, this should not be a significant concern.

11.2.2.3 Water

Again, the interest here is in the volume of water consumed by machining processes. Because water is a relatively low viscosity fluid medium, turbine flowmeters are typically selected for any volumetric flowrate measurement. While other flowmeters, such as the positive displacement flowmeters described for cutting fluid monitoring, can be used for water as well, turbine flowmeters represent an affordable, relatively high accuracy selection ideal for lower viscosity fluids.

Turbine flowmeters operate by placing a small turbine within the fluid flow [2]. The flowrate is then proportional the rotational speed of the blades. Several means can be used to measure the rotational speed of the blades, but a popular choice is to apply the Hall effect to determine the number of complete rotations of the blade. This approach takes advantage of the fact that the metal content of the blade alters the magnetic field generated in between the fork of the blade, which then generates a voltage that can be measured. Another popular choice is the use of radio frequency to determine the number of complete rotations of the blade without inducing a lot of drag forces. In this approach the blade rotation changes the impedance of a radio frequency oscillator circuit, which can then be measured.

Despite the relatively high accuracy, easy use, and low cost of turbine flowmeters, there are several potential disadvantages that must be addressed during the integration of such a sensor. First, bearing selection for the rotor is important to ensure that frictional effects do not affect the accuracy of any measurement [2]. Self-balancing journal bearings made of tungsten carbide, sapphire, or PTFE are typically selected for water applications because it is a clean liquid. These bearings may sufficiently wear, though, such that constant recalibration may be required. Also, the rotor must be placed in the fluid such that drag is minimal, stall does not occur, and the blades cut the fluid perfectly. This requires that the rotor be "twisted" by an appropriate angle, but even then one must be cognizant of changes in flow profile and how they may affect this placement. The flow profile itself can cause further problems as pulsating flows may be over-read due to the inertia of the rotor, and any swirl in the flow may not be captured given the placement of the rotor. To correct these problems, an upstream flow straightener should be used to ensure smooth, constant flow. Lastly, any particulate matter in the flow or high viscosity may damage the rotor and/or bearings. Since machining processes typically use clean, fresh water, though, these problems will likely not affect our measurements.

11.3 Applying Sensor Flows in Decision Making: Automated Monitoring

The sensors described in the previous section can be applied in characterizing the various flows that determine the environmental impact of a manufacturing process. In order to apply the sensors flows in decision making, automated tools are needed to measure and parse the flows [4]. Automation is a vital requirement due to the number of flows involved in a manufacturing process and the complexity involved in applying them in decision making. Moreover, manufacturing systems can be studied at different levels of analysis, ranging from that of the entire enterprise to the tool–chip interface. Each of these levels of analysis also has a corresponding temporal scale of decision making, which ranges from several days at the enterprise level to microseconds at the tool-chip level. Figure 11.2 illustrates the range of variation in the analysis and temporal scales, along with the types of decisions that are made at each level.

Automated monitoring can significantly decrease the complexity of working with large systems. While manual approaches are available for measuring flows (such as with handheld power meters), they can be cumbersome for simple systems, and impossible for more complex systems. A more important consideration is the need to analyze the temporal aspects of the flows. In order to decrease energy consumption, sensor data has to be placed in context of the manufacturing activities.

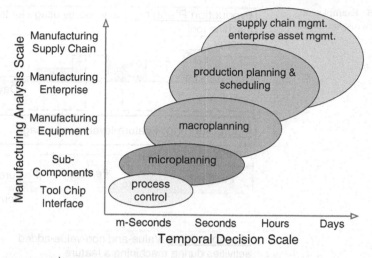

Fig. 11.2 Level of analysis of manufacturing with temporal decision scales

Automated monitoring systems can help attach contextual process-related information to the raw energy data.

Another motivator is the growth of smart grid technologies for energy generation, transmission, and delivery [5] that necessitates the integration of factory flow measurements with the "outside world." The goal of these technologies is to save energy, reduce cost, and increase reliability. Smart energy delivery technology includes demand response, where demand requirements are driven based on energy pricing and availability. While demand response technologies are taking hold in consumer energy markets, their growth has been much slower in industrial markets. A main reason for this has been the lack of detailed demand data from the manufacturing facilities that can be used to drive grid requirements. Moving towards automated monitoring systems will allow better communication of manufacturing system demand data to the grid, enabling smart grid technologies in manufacturing systems.

11.3.1 Approach

Past efforts in sensor-based monitoring and analysis of manufacturing systems have been performed either as an accounting exercise or by using theoretical estimates of the energy required for the various tasks and subtasks involved in manufacturing a part. The former approach is not granular enough to support decision making at the different levels outlined in Fig. 11.2, and the latter approach is not accurate enough, especially in complex systems.

A more effective approach, proposed here is a software-based approach for automated sensor data reasoning, which can support decision making across the

Fig. 11.3 Examples of analysis across temporal scales

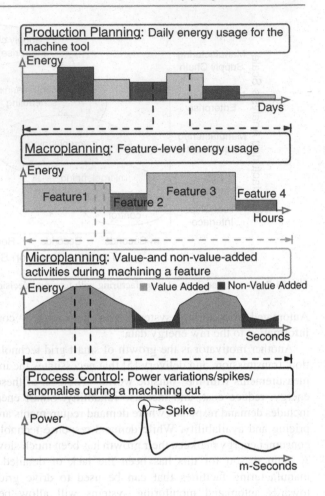

Production Planning: Daily energy usage for the machine tool

Macroplanning: Feature-level energy usage

Microplanning: Value-and non-value-added activities during machining a feature

Process Control: Power variations/spikes/anomalies during a machining cut

multiple temporal levels shown in Fig. 11.2. Figure 11.3 shows examples of the types of analysis required across some of these levels. Based on the analytical complexity of manufacturing processes and systems, the software tools need to have the following capabilities:

- Concurrent monitoring of system flows with process data
- Standardized data sources
- Scalable architecture for large data volumes
- Modular architecture to support analysis across different manufacturing scales

Based on these requirements, the automated sensor-data monitoring system is developed using two key components: an interoperability standard for manufacturing data that can normalize data exchange in the manufacturing system, and a rules engine and complex event processing (CEP) system to handle data reasoning and information processing.

11.3.2 Data Standards: MTConnect

The MTConnect standard for data exchange is selected for data collection from the manufacturing equipment. MTConnect defines a common language and structure for communication in manufacturing equipment and enables interoperability by allowing access to manufacturing data using standardized interfaces [6]. It is an XML-based standard, and describes the structure of manufacturing equipment along with the near-real-time data occurring in the equipment. It allows data from equipment to be logically organized without being constrained by physical data interfaces. Previous work by researchers has used the MTConnect standard in monitoring the energy consumption of machine tools [7]. With MTConnect, the operational data of the machine tool can be monitored in context with the energy consumption data.

11.3.3 Reasoning: Event Stream Processing

Events can be understood as something that occurred either at a point in time or over a range of time. In manufacturing systems, events can be a numerical value (e.g., the instantaneous power consumption at a point in time) or can be a type of annotation (e.g., the alarm state of the machine tool over an interval). Complex events are abstractions of events that are created by combining simple events. For example, a complex event indicating that the machine tool's spindle has crashed can be created based on simple events pertaining to the tool position, the instantaneous power consumption, and the machine tool's alarm state. Event streams are linearly ordered sequences of events and, in this context, correspond primarily to time series events. Events stream processing techniques include rules engines (RE) and complex event processing (CEP). These techniques can be used to create higher-level abstract events and reason on them by pattern matching and identification.

11.3.4 Case Study

The following case study is presented to illustrate the development of an automated system to reason over sensor data flows. It specifically focuses on energy measurements and demonstrates the use of energy monitoring in reasoning over the performance of a manufacturing system. A simulated energy profile is developed based on actual measured data from Toenissen [7] and Diaz et al. [8] from the end-milling of aluminum using a 2-flute carbide cutter in a 3-axis precision milling machine. The simulated profile extrapolates the laboratory measurements to a more representative industrial case with multiple parts/cycles, machine tool events, and longer operational time. The profile also corresponds broadly with the research presented by Hermann et al. [9]. The spindle occupies three states during this profile: idle (0 rpm), low (8,000 rpm) and high (16,000 rpm). The operational states of the machine tool include startup, shutdown, idle, and in-cycle (machining a part). The machine tool demonstrates "spikes" in the energy consumption during a spindle

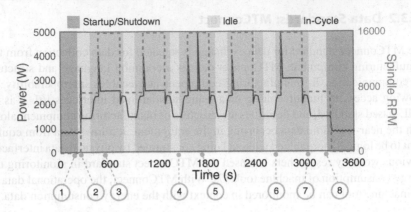

Fig. 11.4 Energy consumption and spindle rpm profile for Case Study

Table 11.2 Event Reasoning Case Study

Event	Time (s)	Reasoning
1. Machine idle	242	Average energy use < idle threshold; spindle speed = 0.
2. Expected energy spike	243	Spike due to spindle start-up (0–8,000 rpm).
3. Expected energy spike	464	Spike due to spindle speed increase (8,000–16,000 rpm).
4. Idle energy constant	1,457	Previous two idle periods energy use constant at 124 kJ.
5. Anomalous spike	1,679	Energy spike unaccompanied by shift in spindle RPM. Potential failure in spindle.
6. Idle energy increase	2,612	Current idle period energy use (211 kJ) > past idle period average energy use (124 kJ).
7. Part energy higher	3,074	Current part energy (1,218 kJ) > previous parts average energy (1,087 kJ).
8. Idle energy trend	3,309	Idle energy increasing monotonically over past two periods (342 kJ > 211 kJ > 124 kJ).

speed change. The power consumption increases when the spindle is engaged at a higher rpm and when there is material removal (or when the machine is in-cycle).

The temporal analysis framework is applied in automatically detecting phenomenon pertaining to relationships between energy usage and operational performance of the machine tool across various temporal ranges. The software framework was implemented using the open source rules engine package, Drools Expert [10]. The specific events identified by the analysis framework with the energy and spindle profiles are shown in Fig. 11.4. The events are discussed in more detail in Table 11.2.

This automated temporal analysis supports decision making to improve the operational and environmental performance of the machine tool. Some scenarios supported by this analysis include:

- Reduce total energy use for the machine tool based on the usage during idle and non-value-added periods.
- Identify disruptions in smooth part production based on anomalous power usage spikes.

- Track maintenance state of the machine tool using historical power usage profiles.
- Enable environmental reporting on a per-part basis by accurately accounting for the energy use of the part as it is being manufactured.
- Notice emerging trends in the energy usage, such as increased total consumption for successive parts, which may indicate process plan deviations and inconsistencies.

11.4 Summary

This chapter looked at various technologies that can be applied in characterizing the "greenness" of manufacturing processes. Quantitative characterization of green manufacturing processes requires sensors to measure and characterize the various flows involved in the manufacturing process. Given the complexity of managing and parsing the sensor data, software tools are needed to automate data monitoring. The chapter presented a framework based on event stream processing to temporally analyze the energy consumption and operational data of machine tools and other manufacturing equipment. This can be expanded to analyze other types of environmentally pertinent data streams in manufacturing systems, which enables decision making to improve the environmental performance of machine tools. As demonstrated, the event stream processing technology enables reasoning of vast data streams over the events that occur in the streams, as well as complex, abstracted events. This capability simplifies environmental analysis and optimization for complex manufacturing systems where decision making is required across multiple levels of abstraction.

References

1. Whitaker J (1998) AC power systems handbook, 2nd edn. CRC, Boca Raton, FL
2. Baker R (2003) An introductory guide to flow measurement. ASME, New York
3. Baker R (2000) Flow measurement handbook: industrial designs, operating principles, performance, and applications. Cambridge University Press, Cambridge
4. Vijayaraghavan A, Dornfeld D (2010) Automated energy monitoring of machine tools. CIRP Ann Manuf Technol 59(1):21–24
5. US Department of Energy (2009) The smart grid: an introduction. http://www.oe.energy.gov/SmartGridIntroduction.htm. Retrieved 20 Feb 2011
6. Vijayaraghavan A, Sobel W, Fox A, Warndorf P, Dornfeld D (2008) Improving machine tool interoperability using standardized interface protocols: MTConnect, Proceedings of ISFA
7. Toenissen S (2009) Power consumption of precision machine tools under varied cutting conditions. LMAS Report. Univ of California, Berkeley
8. Diaz N, Helu M, Jarvis A, Tonissen S, Dornfeld D, Schlosser R (2009) Strategies for minimum energy operation for precision machining. Proceedings of MTTRF, 2009 annual meeting, Shanghai, PRC
9. Hermann C (2009) Simulation based approaches to foster energy efficiency in manufacturing. CIRP GA 2009, CWG EREE Presentation
10. Drools (2009) Drools expert (rule engine). Available at http://www.jboss.org/drools/. Accessed 20 Feb 2011

Concluding Remarks and Observations About the Future

12

David Dornfeld

If you don't know where you're going, you might not get there

Yogi Berra

To succeed, jump as quickly at opportunities as you do at conclusions.

Benjamin Franklin

Abstract

This chapter first looks at the principal developments in manufacturing over the last several hundred years and where this is leading in the future. The role of sustainable manufacturing, and requirements to actually enable this, is discussed. Then, some discussion of "what is manufacturing's role" relative to the life cycle impact and resource consumption of products is presented. Depending on what phase of product represents the most impact will determine the strategies for greening the product. Finally, the concept of "leveraging" manufacturing to insure the maximum improvement in product life cycle impact is introduced.

12.1 Introduction

This book has tried to accomplish a number of things. In no particular order they are help to define, in engineering terms, some of the motivations, challenges, opportunities, tools, and methodologies associated with the discussion of green and sustainable manufacturing, review the ongoing research and educational activities of

D. Dornfeld (✉)
Laboratory for Manufacturing and Sustainability (LMAS), Department of Mechanical Engineering, University of California at Berkeley, 5100A Etcheverry Hall, Mailstop 1740, Berkeley, CA 94720-1740, USA
e-mail: dornfeld@berkeley.edu

D.A. Dornfeld (ed.), *Green Manufacturing: Fundamentals and Applications*,
DOI 10.1007/978-1-4419-6016-0_12, © Springer Science+Business Media New York 2013

the Laboratory for Manufacturing and Sustainability at UC-Berkeley, present some specific examples of the application of these tools and methodologies to challenging manufacturing problems, encourage individuals, companies, and other organizations to think about green manufacturing in the context of something that is achievable in the short, medium, and long term and lay out some ideas for the future directions of green manufacturing. This chapter is focused on this last part—the future.

This chapter starts with the past developments, generally, in manufacturing and follows that trajectory into future directions for manufacturing. First, the principal developments in manufacturing over the last several hundred years are reviewed and comments as to where this is leading in the future are made. The role of sustainable manufacturing, and requirements to actually enable this, is discussed. Then, some discussion of "what is manufacturing's role" relative to the life cycle impact and resource consumption of products is presented. Depending on what phase of product represents the most impact will determine the strategies for greening the product. Finally, the concepts of "leveraging" manufacturing to insure the maximum improvement in product life cycle impact and the energy of labor as a parameter in life-cycle analyses are introduced.

12.2 Evolution of Manufacturing

Society has always affected industrial development and evolution. This extends to manufacturing as well. Technology advances the leveraging of scientific investigation, discovery, and, eventually, reduction to practice. Manufacturing is part of that scientific discovery and reduction to practice. Now the interest in minimizing the impact of manufacturing across the product life cycle is added to the discussion.

To understand where manufacturing has come from and its evolution it is usually helpful to see where technology has been to help determine what the future will hold. This is certainly true for manufacturing and, now, green manufacturing. There are many ways to look at the past, for example, to consider the evolution of "mechanization" of manufacturing and individual processes. This is especially interesting with respect to the relationship of mechanization to the increase in power and control of machines and processes.

Early researchers in manufacturing and the machine tool/manufacturing machinery industry, such as Dr. Eugene Merchant of Cincinnati Milacron company, were early observers of the need to "systematize" machine tools and manufacturing. Driven by an interest in automating manufacturing as well as a pragmatic vision of "100% availability" for machine tools, Merchant proposed an integrated system of machine tools, robots or part handling machinery, tooling and fixturing, and inspection all controlled by central computers. This was in 1983. The idea was computer integrated automation—moment by moment optimization of manufacturing processes and decision making in addition to the control of a large and constantly changing variety of produced items with numerically controlled machines and support hardware.

There was no thought to energy consumption or "green" manufacturing in those days—only increasing productivity and flexibility. Earlier, Henry Ford commented that "reduction of waste (in any form) is desirable" [1]. Ford continued that "…we will not so lightly waste material simply because we can reclaim it—for salvage involves labor. The ideal is to have nothing to salvage." Ford was not known as a leading environmentalist but he knew the value of wasted materials, labor and energy and how to minimize this waste in industrial settings although we might not, now, endorse some of his methods.

Merchant had also observed another key driver to the development of flexible manufacturing systems—social and economic factors. Workers were, even in 1983, increasingly shunning jobs in manufacturing and it was felt that machinery needed to fill the gap. For workers in manufacturing, the industry and government were striving to improve working conditions and automation was the solution. Tie this into concepts of lean manufacturing and *kan ban* and the "future of manufacturing" was conceived. To a great extent the world is still trying to build that future. One big difference is that the information technology of today allows things that Merchant could only have dreamed about.

When looking at the future of manufacturing—or, at least, about what will be influencing the world in the future and to what extent this will drive green technology and sustainable development one can consider the individual components of the triple bottom line—social values, economic values, and environmental values. The basis for assessing environmental values is well established and there have been examples presented earlier in this book. Society and economy need a bit more discussion—especially with respect to future directions and drivers.

12.2.1 Society

Society has always affected how industry develops and evolves. This is more apparent today as society finds itself challenged by problems that cannot be solved only by technology—for example, global warming and the access to energy, water, materials, etc. Sustainable development (and all of the requirements and impacts that will entail) will determine many aspects of the future of manufacturing. The three elements (or stakeholders) shown earlier in Fig. 1.1 must each be considered equally as a sustainable manufacturing base for living is built. The "sweet spot" for sustainability or a sustainable business is right in the center at the intersection of the three elements—a balance of social impact and needs, environmental impacts and needs, and business and economy impact and needs.

There are many definitions of sustainability that have been cited earlier in this book but, in general, the idea is that one should not do anything today that would limit the well-being or quality of life of future generations. Here well-being would have as a basic requirement access to clean air, water, sufficient food, and shelter and opportunity for each individual to pursue whatever the individual wishes as a career path. Right now this is not true for many of the people around the world. A balance of social needs (education, employment, health care, nourishment, the rights of the individual, etc.), the environment (sustainable use of natural resources, water, renewable energy, protection of nature, etc.) and the economy (business success, free enterprise, distribution of wealth, infrastructure, etc.) will require a different view of the value of all these elements in the world of the future. It also means that if one aspect of life is not sustainable (say energy use, for example, due to energy use per person increasing as well as the number of people in the world increasing) the technological developments must drive reduction of energy use to offset the initial degree of overuse/person as well as enough to accommodate the increasing population! This will mean big changes to our way of business if this is to be realized. And this will mean big changes to manufacturing and manufacturing machinery as well. But, they offer part of the solution.

And there is a strong and growing link between business and society.

12.2.2 Business and Economy

The outlook proposed in the section above requires a dramatically different view of the economy as a start. Lester Brown observed in The Second Coming of Copernicus [2] that

Today, we need a shift in how we think about the relationship between the earth and the economy. This shift is no less fundamental than the one proposed by Copernicus back in 1575. This time, the issue is not which celestial sphere revolves around the other—but whether the environment is part of the economy—or the economy is part of the environment.

Chapter 1 and subsequent examples throughout this book outlined the pivotal role that manufacturing plays in the life cycle of a product—products upon which businesses depend for their livelihood and upon which the economy is built. The increasing costs of energy and other raw materials, the reduced availability of many vital resources, the impact of industry and consumption on the environment and the awakening of the public to the need to and desire for change will drive our manufacturing future. The economy is indeed part of the environment.

Business has been concerned about sustainability and the impacts of humans on the environment for some time. The current challenges heighten our awareness of the need to take action systematically. The Ricoh Company in Japan introduced what they called the "comet circle™" in the late 1990s to graphically represent a society that recirculates resources. Industry plays a large role in this. According to the circle concept, the closer to the consumer that the recycling or reuse occurs, according to the circle, the more efficient the use of resources is. Logically, as you move away from reuse or recycling close to the consumer, more processing is involved, more transportation, more loss of energy and materials, degradation of material properties, etc. can occur. This puts a lot of responsibility on the manufacturer to create products that can be reused or recycled effectively, or upgraded and returned to the consumer without needing to start back at the resource extraction stage. This will take time. Changes in product design, whether it is a coffee maker, computer, aircraft, or solar panel, or machine tool, to address the reuse and recycle challenges will evolve over time. And, in use, the products will be required to efficiently use energy and other resources.

Similarly, manufacturing will be required to embody these changes. The machinery and hardware in a factory (where ever it is in the world and whatever role is plays in the supply chain) derives from the same cradle-to-cradle (or cradle to grave) life cycle. Machine tools are products. They will need to be manufactured using sustainable technologies, operate to use minimum energy and resources, be able to be reused or recycled effectively at their end of life. And the factories that build machine tools, like the homes people live in and the other facilities that provide the products and services everyone depends on, will need to operate with minimum energy and resource use and will minimum or zero waste or impact to the environment. That means these factories will also have to operate according to the comet circle. This emphasizes the importance of renewable energy and renewable resources to meet the goal of sustainable manufacturing.

To determine whether or not this kind of development and progress is possible, it is necessary only to look at how far industry and manufacturing have come over the past 200 years. If one reviews the progress, or evolution, of manufacturing in terms productivity, flexibility, response time, work philosophy or business model, or market responsiveness/customer "pull," one can see tremendous changes from the earliest organized industry or manufacturing in the 1800 s to today.

These changes occurred in specific "leaps" of development/progress:

- Craft production: In the early days of industry-skilled workers, artisans, worked on a variety of machines to create special build products with high labor input. Very little mechanization was available to the craftsman, who was completely on his own in terms of process planning, timing, and techniques used.
- Mass production: In the late 1800 s to early 1900s saw the development of mass production (e.g., Eli Whitney, Henry Ford and the assembly line, etc.). This yielded a dramatic reduction of direct labor (at the expense of the craftsman), higher production rates, more control of the process, ability to satisfy larger customer demand, interchangeable parts, and the first elements of automation. Cost per piece dropped substantially.
- Flexible production: In the 1980s, thanks to visionary thinkers like Eugene Merchant in the USA, Japanese manufacturers like Taiichi Ono and the Toyota Production System, and others, a tremendous increase in the efficiency of these production systems was seen. Unnecessary operations were done outside of the manufacturing process on the machine resulting in high levels of machine utilization thanks, in part, to the computer and clever methods of preparing the workpieces and tooling for the machine off line while assuring quality and accuracy.
- Lean Manufacturing and "Mass Personalization": The late 1990s and 2000 saw the introduction of concepts of true response to customer demands with abilities to manufacture customized products in small quantities with the efficiency of mass production and with short lead times (order placed to product delivered). This was enabled by strict quality methods to insure "first part correct" built on Ohno's ideas and the strict quality control methodology taught by W. Edwards

Deming and others thus eliminating backup stock, inventory (and the cost of capital associated with mounds of parts and products kept available to cover for manufacturing faults or inability to plan for or respond to customer demand).

Reviewing all these "leaps" one asks the question—what caused these big shifts between each of these phases? Each change occurred because of a realization that an improved system of manufacturing could be realized if the system was "designed and optimized" based on an understanding of some new criteria—more control of the process and standardization introduced by Henry Ford, better use of manufacturing machinery and increased productivity introduced by Taiichi Ohno, and reduced inventory and buffer stocks for quality production of Edwards Deming and Ohno. That is, costs and other considerations that had been previously included as a normal part of the process, that is, internalized, were externalized, counted and either avoided, deleted or put in parallel to the core process.

12.2.3 The Next Big Change

Following these improvements, then, what is expected to be the next big change? This is relatively easy to predict—sustainable production. How will this be accomplished? The driver for this big change is to include the true cost of the production of a product from resource extraction to end of life and reuse or recycling in the cost of the product. This true cost is more than the "value added" through all these stages (which one might argue is already included following the best of capitalism). We need to add the environmental and social costs associated with the life cycle. This will not be easy and it is not proposed that all the information or tools necessary to do this exist today. It is necessary to incorporate a cost for all the embedded energy, materials and other resources, labor, impacts on the environment, and accompanying social requirements and impacts, etc. in the cost of the product. Then, when the consumer purchases a new computer, automobile, airplane ticket, machine tool, etc. she will "see" the true impact of that product reflected in the price. And, on that basis, can shop around. The cost of recycling, or disposal, now covered by local governments, or whoever is paid to pick up the trash at our curbs every week, will be reflected in the cost of the product.

Manufacturers will design and build products to minimize this "life cycle cost" just as they did to exploit mass production, or lean manufacturing, in the past. The comet circle™ of Ricoh will be a roadmap to minimizing product cost for sustainable manufacturing. The unit of measure of economic value may well be the "carbon footprint" of the product (i.e., a measure of how much carbon, greenhouse gases, or equivalent) was used in the manufacture and operation of the product—how much of an imprint on the world does the manufacture and use of this product create? Distributed manufacturing with "world-wide factories" was based on relatively inexpensive labor in other parts of the world and very inexpensive transportation to move products to market. Not surprisingly, as countries develop the cost of labor inches upward. In addition, as fuel prices rise (especially faster than normal inflation would dictate) the cost of transportation can become unacceptably high.

Then the model of building components of a product in facilities scattered around the world and assembling the product in one of the locations for shipment to another location may not serve business well. If life-cycle costs are considered, the entire supply chain might look very different and manufacturing in locations of clean, renewable energy, water supplies, and reasonable labor costs may offset the high cost of transportation and penalties assessed to products built where the electricity generated contributes more greenhouse gas emission per kilowatt hour due to coal or other carbon-based sources of energy for the power plant.

Engineers will need to be educated in the tools and metrics for assessing the impact of a product and whether or not it meets the requirements of sustainability to be able to function in this industry. This will be integrated in the product development cycle and in the computer aided design and manufacturing tools that engineers use to realize the design and manufacturing procedures. And business and government will need to see the benefit of sustainable manufacturing. Importantly, there will not be one "silver bullet," one solution to address all these challenges. Rather, it will be the accumulated impact of many, many technological, social and business advances, wedges if you will, together that finally move towards a sustainable world.

12.3 Leveraging Manufacturing

12.3.1 Background on Leveraging

Manufacturing offers many opportunities for reducing environmental impact, utilizing resources more efficiently and, overall, greening the technology of production. These opportunities are most often related to process, machine or system improvements that impact only the operation of the process, machine or system. But, there is more potential in manufacturing enhancements to have a larger impact on the life cycle impact of the product the manufactured item is used in. This is referred to as "leveraging" and identifies manufacturing-based efficiencies in the product that are due to improved manufacturing capability but which, in the long run, have their biggest effects on the lifetime consumption of energy or other resources or environmental impacts.

First, what is meant by the term "leveraging"? A lever is understood to be a device to increase mechanical advantage, as a bar used with a fulcrum to pry a heavy load allowing a larger load to be moved than with simple force alone. Leveraging is used as a transitive verb, usually in financial discussions such as [3]:

The use of credit or borrowed funds to improve one's speculative capacity and increase the rate of return from an investment.

The general idea is to employ resources in such a way as to insure a larger return on the effort (or in financial terms, money) than might otherwise be realized.

How this relates to manufacturing, and, in specific, green manufacturing will depend on the component being manufactured by a machine or process and its eventual use in a product. It will also depend on where the largest impact of the product comes from—the manufacturing phase or the use phase.

Fig. 12.2 Use vs. manufacturing phase resource intensity and impact [4]. Copyright 2007 LMAS reprinted with permission

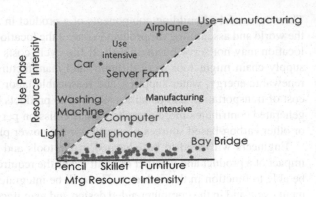

12.3.2 Use vs. Manufacturing Phase Impacts

Depending a lot on the product and its use, different impacts (meaning consumption of resources or environmental impact resulting from that consumption) can occur. This can be illustrated in Fig. 12.2 [4] and plots use phase resource intensity as a function of manufacturing phase resource intensity.

In the figure "things that don't move or need power to operate" like bridges, furniture, etc. are dominantly manufacturing phase consumers of resources and, by extension, impact. Things that do "move and need power to operate" like automobiles, airplanes, etc. are use phase heavy. Interesting to note are the items that are close to the break-even 45 degree line. Personal computers overall, but not the chips in them, are a bit heavier in the manufacturing phase than use phase. Cell phones more heavy (but likely not if you include the embedded impact of the infrastructure needed to operate a cell phone network.) As usual, the details matter.

12.3.3 Strategies for Product Greening

One can actually visualize this use vs. manufacturing impact space in terms of what needs to be done to improve product impact (or performance) depending on where the product sits in that space. One can identify four quadrants of "sustainable product" characterization, Fig. 12.3. The axes are the same as in the use vs. manufacturing discussion and indicate, from low to high, the consumption or impact of that phase of the product's life cycle. Then the "low–low" quadrant indicates the most sustainable product. The "high–high" quadrant contains products that are to be avoided or, in another sense, offer the most potential for improvement. The two "high–low" quadrants represent products where we need either to increase the efficiency of the product (with respect to design or using manufacturing leveraging) or to improve the efficiency of the manufacturing process relative to use and manufacturing phases, respectively.

Fig. 12.3 Regions of product performance

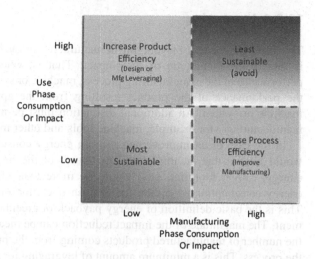

Depending on where a product lies in this diagram, and assuming it is not yet in the lower left quadrant of "most sustainable," one can determine the strategy for improving product performance or manufacturing impact. For example, if the product is in the "high manufacturing phase consumption/impact and low use phase consumption/impact," that is the lower right quadrant, then it is incumbent upon manufacturing engineers to reduce the manufacturing impact. This could be done by a number of means. If the process is a "high tare" process (meaning the manufacturing process uses almost as much energy or resources when in idle mode as it does in production—for example a heat treating furnace or a high precision machine tool) then one should look for ways to increase production to minimize the per part energy or resource content. For a process that is highly energy intensive one needs to modify, or replace, the process with an alternate process. Or, one can look at combining operations to eliminate energy intensive process steps.

Alternatively, if the product is in the "low manufacturing phase consumption/ impact and high use phase consumption/impact" quadrant, that is, the upper left quadrant, then it is incumbent upon design engineers to reduce the use phase impact. This could be done by increasing the efficiency of the product (i.e., better use of energy in operation) or reduce the impact by reducing the waste or non-useful output of the process. Recent improvement in automobile fuel economy is an example of increasing the efficiency of the product. This will add up significantly over the product's lifetime.

The question is raised, what if manufacturing is a small contributor to the life cycle impact of the product (either in terms of consumption or impact)? Is it then that engineers ignore manufacturing improvements and rely only on design of product to improve performance? This is where "leveraging" can be valuable.

12.3.4 Leveraging

Two different classes of leveraging of manufacturing can be defined. The difference is due to the magnitude of the impact. That is, whether it impacts only the performance of the manufacturing process, machine or system or whether it impacts the performance of the product resulting from the application of the process, machine or system. An additional distinction must be made for products used in manufacturing—for example, machine tools and other manufacturing hardware.

In the case of an improvement, say in energy consumption of a process, one would require that, at minimum, the "cost" of the improvement (in embedded energy, carbon footprint, etc.) would be more than offset but the reduction in energy consumption or carbon footprint in operation of the "improved" process. This is the basic definition of energy payback or greenhouse gas return on investment. The magnitude of the impact reduction can be measured simply by knowing the number of manufactured products coming from the process over the lifetime of the process. This is a minimum amount of leveraging for any contemplated process improvement to insure that we are making progress.

A second, more impressive, leveraging is due to process (or machine or system) improvements that have an inordinately high ability to reduce the impact of the product of the manufacturing operation (or machine or system) over the lifetime of the product use. The original process improvement may not have been made as part of a greening analysis of the process but is due to the introduction of new technology, machine capability, or materials. It is this second type of leveraging that is likely to have the greatest potential for reducing the T term in the impact equation—making a larger than normal reduction in the product impact/GDP during the product's life time.

Why this distinction is important is discussed in the next section.

12.3.5 Does Manufacturing Matter?

The base of this discussion is an assessment of whether or not manufacturing is a significant component of energy and resource consumption and the impact from this consumption, and, then, whether or not changes in manufacturing can really help overall. A review of all the data, pie charts and discussions about how much of the world's energy use is attributed to manufacturing is not presented here.

Allwood [5] points out that industrial carbon emissions are predominately due to production of goods in steel, cement, plastic, paper, and aluminum. With the demand for these materials expected to double at least by 2050, during which the global carbon emissions are desired to be reduced by at least 50%, simply improving process efficiency will fall far short. Allwood suggests several strategies for industrial emissions reduction in addition to process efficiency, increased recycling, and carbon sequestration and storage, as (1) reducing demand for materials; (2) nondestructive recycling; and (3) radical process innovations which allow shorter, less energy intensive process routes to yield the completed component.

But, for "general product manufacturing" is there enough that can be accomplished by manufacturing improvement? Especially if this is applied to what many manufacturing operations consider core process capabilities—like machining?

It is helpful to consider a specific example. In the discussion about use vs. manufacturing phase impacts the automobile has a distinctly "use phase" impact. At a presentation at the ICMC Conference in Chemnitz in September 2010 by a representative of a large German automaker the speaker mentioned that, by their analysis, about 20% of the impact of a typical small sedan car they produce came from manufacturing while 80% was due to the use phase.

Materials and part suppliers account for much of the embedded energy in the manufacturing phase. Machined components, such as the gear box and engine are a small percentage of the total (accounting for about 10% overall or about 25% with materials and parts from suppliers included).

If one looks at the impact of the auto, including car production, fuel production and use phase it is clear that the fuel production and consumption in the use phase dominates all categories of emissions to air and water with the exception of dust generated by material production and casting of some components and painting of the vehicle and biological oxygen demand impacts on water [6].

Thinking a bit more about the data above, does this make sense in terms of reducing the impact/GDP? If the focus is only on the manufacturing phase it is not very encouraging—especially if the predominant impact is in the use phase.

Consider the automotive example of 20% manufacturing phase impact vs. 80% use phase impact above. Assuming that about 20% of the manufacturing is machining or machining related, that gives a potential for an improvement of 20% of 20% or only 4% (and then only if all machining is eliminated—not likely). Assume that some of the better technology for improving machining efficiency is employed, say some specialty tooling material that reduces machining power consumption and that is worth another 20%. Now the impact of machining as part of manufacturing on the life cycle impact of the auto is down to 0.8% (20% of 4%).

One could argue that this is hardly worth the effort it would seem. Of course, for the company paying the electricity bill for the if this 0.8% technology wedge is added to a lot of others in machine operation it can add up to real savings. But, still not impressive compared to use phase impacts. That is, impact over the full life cycle of the auto.

12.3.6 Accounting for More of Manufacturing's Impact

The question is, then, what is the true leverage effect of manufacturing on the life cycle impact of a consumer product—one that has its dominant impact in the use phase rather than the manufacturing phase? If one is referring to a manufacturing machinery builder, like a machine tool company, then one can argue that the machine tool has its largest impact in the use phase so that improvements in energy efficiency of the machine will been seen over its life since it is the "product" [7].

The thesis here is simple. If improvements in manufacturing yield a substantial reduction in the life cycle impact of a product, should not manufacturing get some of the "credit" for this improvement? And, by similar reasoning, can one claim this as a part of "green manufacturing" contribution towards sustainability since it is a major element in reducing the technology impact of the product?

An example of leveraging is instructive. Manufacturing has a number of fundamental effects on a product. In no particular order, manufacturing can:

– Guarantee a certain level of precision or accuracy of the produced component.
– Allow the use of advanced materials (enhance strength to weight, improved surfaces, wear resistance, thermal stability, etc.).
– Allow reductions in process steps or sequences.
– Combine processes for enhanced effects as in hybrid processes or mill-turn machine tools.
– Achieve complex shapes or features to improve performance and so on.

How these manufacturing-induced effects influence the life-cycle performance of the product must be clearly understood to explain the full potential of leveraging. This influence usually comes from the extension of one of the above listed effects onto the energy consumption or "environmental performance" of the product the manufactured components are used in.

A simple example might be a spindle motor for a machine tool. If the production technique for the motor, using advanced magnetic materials, allows the construction of a motor that extracts more useful work from the energy supplied to it, then the manufacturing effect is leveraged over the life of the spindle.

Alternatively, if the improvement in energy consumption is due to controller related performance enhancement, as the 40% reduction in energy consumption illustrated by Mori Seiki due to overall system component improvements and optimum acceleration of spindle and servo motor during machining [8], this is not due to manufacturing leveraging but, certainly, improves the life cycle impact of the machine tool—the product in this case.

Two examples are presented here that illustrate the concept of leveraging manufacturing with life cycle impacts on the product that the manufactured component(s) is (are) used in. And the life cycle impact is substantial and most of the benefits are due to manufacturing.

Both of these examples relate to improved machining tolerances and their impact on product performance. On an aircraft airframe (a large one like a B747 or the A380) savings in weight correspond directly to savings in fuel. And many other aspects of an aircraft scale with weight. This is, to some extent, true also for an automobile. That is the second example.

If the machining process for large airframe components is under control and precision manufacturing principles applied, a reduction in machining tolerances from approximately ±150 μm to ±100 μm on the features of the airframe can account for a weight reduction of 4,500 kg/aircraft and substantial fuel savings (8%) [9]. This allows an increase of 10% in passenger load (the engines don't need to carry as much plane), or increase in cargo payload and a substantial reduction in manufacturing cost of the aircraft (less material and improved assembly) and the

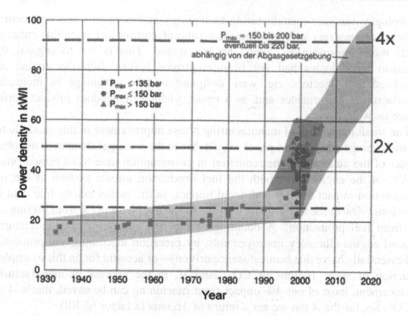

Fig. 12.4 After [10] Change in power density over time for Diesel engines

accompanying reduction in scrap. And less fuel consumption means reduced CO_2 impact from aircraft operation. The accumulated savings over the life of the aircraft are incredible. The fuel consumption per km is estimated as 11.88 L/km (or about 5 gallon per mile). Thus, the CO_2 emission rate can be estimated at 30.64 kg/km. A reduction in fuel consumption of 8% results in a reduction of almost 2.5 kg/km CO_2. And this is over the life of the aircraft—many millions of kilometers.

The next example relates to a similar impact on product use for an automobile. It is also due to enhancement in manufacturing capability due to precision manufacturing.

The improvement for the Boeing aircraft example was based on tightened tolerances allowing increased structural performance by better control on dimensions—resulting in lower weight components. Looking at improvements in engine performance for automobiles we can see similar improvements. The performance (power density in kW/L) of diesel passenger car engines is shown in the graph in Fig. 12.4 from [10]. With better tolerances, better surface finishes, better control of orifice size, and shape on the fuel injector nozzles (with diameters on the order of 60 µm), tighter control on cooling channels and fluid flow in the engine due to enhanced casting techniques, and so forth, the engine (still working on the same old Diesel principles) performs dramatically better.

The "dog leg" in the chart above corresponds to the introduction of high performance, precision, manufacturing to the power train manufacturing in the automobile. In the years since 2000, the power density has been improved by double (in 2007) and anticipated to quadruple by 2020; Similar improvements can be seen in the transmission as well. And, with advanced sheet metal forming technologies (another manufacturing technology enhancement) and replacement of metal components with non-metallics (manufacturing and materials enhancement) more improvements would be anticipated. This is not to suggest that precision technologies had not been employed before. But, the engine and associated fuel injectors, etc. were designed to take advantage of increasing manufacturing performance and, as a result, yielded tremendous product performance as well.

The small percentage of manufacturing phase improvement in this example has a giant leverage effect on use phase impact. Since the principal element in use phase impact of the automobile, the reduction in consumption (due to increased power density of the engine), hits both the fuel production impact as well as the fuel consumption impact there is additional impact. In the earlier automobile data for emissions, 90% of the CO_2 impact was due to the use phase (81% from driving and 9% from fuel production). A doubling of the fuel economy, by manufacturing induced engine efficiency improvements, by precision machining and processing will essentially halve that (same distance driven)—or account for, in this example, a reduction of some 16 tons of CO_2. And if, in the process of manufacturing enhancement, most of our 4% impact from machining can be saved, that's .4 ton of CO_2. So, for the .4 ton we get a return of 16 tons (a factor of 40!).

Actually, more likely it will require manufacturing processes that will result in an increase in impact (meaning machines that consume more energy or processes requiring additional resources) so that it would be better to calculate the "return in investment" of the improvement. In any case, the leveraging effect is impressive.

There are constraints of course. As noted, the technology enhancement (the "wedge") needed to improve precision of the machine tool to enable some of the product performance increases may not be strictly "green" (meaning there is a cost in terms of embedded energy, energy/unit product, or other measure). Trends in machine and process design are showing that one can enhance the performance of the manufacturing process and also realize reduced impacts. But, this needs to be carefully accounted for.

A second issue is whether or not manufacturing can rightfully claim credit for any or all of these improvements under leveraging. Traditional design textbooks outline the design process in stages with clever designs being turned into real products through manufacturing. So, for sure, the role of manufacturing as a design enabler is undisputed. In that case, we can claim the benefits of leveraging manufacturing as well.

There are other considerations for the future of green manufacturing also. One dominant issue is the value of labor in manufacturing. This will be important as it addresses social issues of course and also can have an impact on the potential for reuse or remanufacturing of products. Labor and energy are addressed in the next section.

12.4 Energy of Labor

12.4.1 Introduction

Finally, what is the role of labor in manufacturing and what role might it play in the future? If a company reduces the amount of machinery used in manufacturing and replaces that machinery with manual labor does that help? The response was typically academic—it's complicated. Clearly there are some products and processes that don't lend themselves to this "conversation." But, for assembly tasks, one might make the argument that more human labor (replacing automation) might produce the product using less energy and resources and, ultimately, making the product easier to disassemble at its end of life. And then one must consider the quality of the labor (meaning is it dull and repetitive or intellectually stimulating and, for sure, is it free from danger or other safety issues). This discussion is based on a research paper and thesis [11, 12].

It is conceivable that understanding this "tradeoff" might define a new economic basis of comparison that could actually encourage improved workplace environments for manual workers.

Energy is an important metric of environmental impact and manufacturing efficiency. It is known that in life-cycle assessment (LCA) analyses, energy consumption as a key parameter that can dominate environmental impacts such as global warming potential, carcinogenic emissions, and acidification potential. Energy assessment is also effective as an indicator of manufacturing efficiency. As yield, manufacturing cycle efficiency, process capability, and other manufacturing performance metrics improve energy use per unit output decreases accordingly.

The metric of energy use was popularized largely due to the work of Howard Odum, who has written numerous books on energy and environmental accounting since the 1970s (e.g., Odum [13]). In another publication, Odum [14] presented several methods of quantifying the energy use of labor, in terms of metabolic energy, national fuel share, national energy share, and as a function of the level of education enjoyed by a worker. Others have also discussed the energy use of labor in the form of caloric content of food consumed. Calculated as such, the conclusion is that the energy contribution of human labor to energy use is negligible. But that cannot be the full story.

Energy of labor and EIO-LCA should not be applied at the same level of analysis because many sources of energy use would be double counted. However, energy of labor can be very effective if incorporated into hybrid process-based EIO-LCA, for example, to assess activity upstream of the process-based analysis. The energy use of labor enriches the horizontal scope of process-based LCA, while EIO captures vertical supply chain impacts.

In addition to improving the accuracy of LCA, evaluating the energy of labor can be applied to extend the decision making capabilities of LCA. The energy of labor enables the analyst to quantify and inform decisions that introduce or reduce the degree of automation, deal with the location of a plant, or involve labor intensive process steps.

The energy use of labor helps address the disparities between environmental and economic accounting. Environmental analysis largely ignores labor, while the cost of labor factors very heavily into economic analysis. Evaluating the energy use of labor can help reduce the gap between those who prioritize environment and those who prioritize economics.

Finally, human capital, like environmental capital, has externalities that can be passed from a manufacturing system to society at large. For example, manufacturers who pay workers less than a livable wage rely on social programs to support their workforce. The energy use of labor is a tool with which one can begin to account for the environmental externalities of labor.

One of the questions that comes out of considering the energy of labor is "could this justify exporting all labor intensive industry" to "make the numbers look good?!" If it appears that manufacturing that is more labor intensive than automation intensive uses less energy/unit of product, one might argue this is a convenient way to offset high energy consumption (and its impact) of domestic production. That would be an unfortunate, and unjustifiable, conclusion.

Without quantifying the energy use of labor, it is easy to underestimate the environmental impacts of labor intensive processes, such as those used in product installation, maintenance, repair, and recycling. For example, energy payback time analyses for solar cells often do not consider panel installation, even though it is a major component of their financial cost. Evaluating the energy use of labor is necessary to determine the impact of expensive and labor-intensive solar cell installation on energy payback time.

Labor-intensive sorting processes for recycling are another important application of the energy use of labor. It is important to know the degree to which the energy expended in sorting processes counteracts the energy savings of recycling. There many benefits to recycling outside of energy savings, but the ratio of energy inputs, including that of labor, to energy savings can serve as a measure of efficiency for recycling operations.

The degree of labor required between industries can vary dramatically. Agriculture, handcraft, textile, and service industries are especially labor-intensive. These industries have typically not been the subject of life-cycle analysis, even though their products are consumed in relatively large quantities. Process-based LCA would in fact grossly underreport the environmental costs of a service or an entirely handmade product.

It is also interesting to note that new industries, such as the renewable energy and nanotechnology industries, typically employ more workers per unit output than more established industries. Emerging industries may present problems for LCA practitioners seeking to perform comprehensive assessments. As EIO-LCA data is not yet available for the industry in question, new technologies must be assessed using process-based or hybrid EIO-LCA. Evaluating the energy use of labor is therefore especially valuable to accurately assess the environmental impacts of new technologies and industries.

12.5 Closing Comments

This book has tried to give a perspective to a complex concept of green manufacturing—how it is defined, where is "fits in" relative to sustainable production, what are the basic guidelines and tools of the trade for analyzing and practicing green manufacturing and some examples of applications. The content reflects the research activities of the Laboratory for Manufacturing and Sustainability at the University of California-Berkeley and is presented in the voice of number of the student researchers in the lab over the last few years. As this is being written, much more has been accomplished recently that will not be included here due to the continuously evolving nature of the subject and the constant advance of research and understanding. Like painting the Golden Gate Bridge across San Francisco Bay, when one finishes at one end it is time to start again from the beginning!

The authors hope that this view into green manufacturing from our perspective will help the reader to understand a bit more some of the practical aspects of the topic, encourage them to look more closely at the processes and systems around them in their work or research to observe the opportunities for improvement, replacement, reuse and, overall, reduction of impact, allow them to quantify the effects of these improvements and impact reductions and, hence, support the overall goal of greening manufacturing as part of the move towards sustainable manufacturing.

References

1. Brown L (2001) The second coming of Copernicus. The Globalist.
2. Ford H, Crowther S (1922) My life and work. Garden City Publishing Company Inc., New York
3. Free Dictionary (2011) http://www.thefreedictionary.com/leveraging. Accessed 16 Jan 2011
4. Zhang T (2007) Laboratory for manufacturing and sustainability presentation. Univ of California Berkeley, Berkeley, CA
5. Allwood J, Cullen J, Milford R (2010) Options for achieving a 50% cut in industrial carbon emissions by 2050. Environ Sci Technol 44(6):1888–1894
6. Schweimer G, Levin M (2011) Life Cycle Inventory for the Golf A4. Available www.volkswagenag.com/.../Golf_A4__Life_Cycle_Inventory.../golfa4_english.pdf. Accessed 18 Jan 2011
7. Diaz N, Choi S, Helu M, Chen Y, Jayanathan S, Yasui Y, Kong D, Pavanaskar S, Dornfeld D (2010) Machine tool design and operation strategies for green manufacturing. Proceedings of the fourth CIRP international conference on high performance cutting, Gifu, Japan
8. Mori M (2010) Power consumption reduction of machine tools. Presentation at 2010 CIRP General Assembly CWG-EREE, August, Pisa
9. Thompson D (1995) Presentation at symposium on research issues in precision manufacturing. Univ of California Berkeley, September
10. Berger K (2005) Daimler. WG on burr formation presentation. CIRP January 2005 meeting, Paris
11. Zhang T (2007) Energy use per worker-hour: evaluating the contribution of labor to manufacturing energy use. Proceedings of the 14th CIRP international conference on life cycle engineering, Japan

12. Zhang T (2011) Producer-focused life cycle analysis of thin film silicon PV systems. PhD thesis, Mech Eng, Univ of California, Berkeley
13. Odum H (1971) Environment, power, and society. Wiley-Interscience, New York
14. Odum H (1996) Environmental accounting: energy and environmental decision making. Wiley, New York

Index

D.A. Dornfeld (ed.), *Green Manufacturing: Fundamentals and Applications*,
DOI 10.1007/978-1-4419-6016-0, © Springer Science+Business Media New York 2013